Carol Klein

GROW YOUR OWN GARDEN

Photographs by
Jonathan Buckley

Carol Klein

GROW YOUR OWN GARDEN

How to propagate all your own plants

Photographs by
Jonathan Buckley

BOOKS

Contents

INTRODUCTION

Growing your own garden is not a novel concept. It's not as though the idea has just been invented, or suddenly become ultra-fashionable. It is real, practical and founded on observing, respecting and emulating nature. We all remember the first seeds we sowed and the thrill we got when the tiny green shoots began to appear. Who didn't run back to have another look to check there was no mistake? Who didn't smile?

I remember clearing a bit of ground soon after we moved here and putting in peas for the first time – one of my earliest ventures into proper seed-sowing – and soon afterwards being very disheartened when they didn't come up. Close to giving up, I asked my then next-door neighbour Charles, a gardening vicar, to come and tell me where I was going wrong. But when we approached the supposedly barren patch, there in perfect symmetry was a matrix of green sprouts pushing their way through the soil!

Long before that, when I was a little girl, my mum put a few wrinkled nasturtium seeds into my hand and encouraged me to push them into the soil in our Manchester back yard. A couple of weeks later we went out together to marvel at the twin round leaves, newly emerged. At the end of the summer we collected seed from the plants and squirrelled it away in brown paper bags – I could hardly wait for the next year. Big seeds like peas

and nasturtiums are tailor-made for introducing children to propagation and helping them feel at one with the cycle of growth. The interest that every child has in growing plants should be kindled from the start.

It is often not until much later, when we have our own plot, that we try our hand at some of the less obvious propagation techniques. I took my first proper cuttings soon after we came here, prompted by a yearning to grow the plants of my dreams. The garden at Glebe Cottage is packed with plants that we have propagated here. There is a wondrous continuity to this, taking cuttings from plants you grew from cuttings, saving seed from plants originally grown from seed, and swapping seed and plants with other gardeners. I have been lucky enough to have had advice from some of the most skilful propagators around, people who do it for a living. And few people do this for a living without loving it.

Propagating your own plants may not be a new concept, but its relevance to 'up-to-the-minute' gardening cannot be over-stressed. Regenerating your plants and growing your own garden is the way forward; it is sustainable gardening *par excellence*. At the same time it is enormously exciting, hugely rewarding, and each time you do it – whether you grow a plant from seed or help a stem to take root – you re-affirm the primacy of the earth and the plants that grow in it.

SEEDS

SNIPPING STEMS
LEFT The time to
collect seed like
this *Melica altissima*
is just at the point
when it is ready to
disperse, and always
on a dry, sunny day.

Whenever I pause to admire flowers in their full glory in the garden, I always have an ulterior motive. Lurking in the back of my mind is the question of how long it will be before those petals fall and are replaced with fat seed pods or feathery heads full of the potential for new life and new plants; how soon can the next revolution begin?

Growing plants from seed is one of the most basic and fulfilling of all gardening activities, and one that humans have practised since we first made the decision to settle in one place and grow our own food, rather than simply collecting or killing it. When we collect our own seed, clean it, dry it, sow it and nurture our young plants, we participate in an endeavour that stretches back for thousands of years. This annual renaissance is synonymous with the cycle of life: it links us to all the previous generations who have collected and grown on their own seeds, and to all the future generations who will do the same.

Although it is rewarding to sow from the bright packets on the garden-centre shelves, how much more satisfying it is to collect your own seed and start a new dynasty of plants! Collecting your own is always better than buying seeds that may have been stored for some

time, and in some cases entered a dormant state. Throughout autumn, as the brown paper bags of precious booty accumulate, the sense of anticipation starts to build. During the bleakness of winter, as seeds are cleaned and packeted, the names scribbled on the bags summon up pictures of the plants that will eventually emerge. You can almost feel the satin of their petals, the soft or shiny texture of their leaves, nearly smell the perfume of their flowers and see their quivering stamens.

This is as close as gardening gets to the earth's heartbeat. No matter how many times you sow seeds, the process is just as magical. As you cast them across a patch of ground or a seed tray, spread them out, cover them and water them, your belief that they will germinate is an act of faith. Whether this is the first or the thousandth time you have done it, as each seed sends out its first inquiring leaves and joins together with its fellows to create a green haze, you always feel the same sense of exultation.

Growing from seed brings out all our nurturing instincts: once you've sown these seeds, you are responsible for them. You will guide them and tend them through infancy and adolescence until they're strong enough to take their place in the big, wide world.

Seed dispersal

Growing from seed is the method of propagation that most closely emulates nature.

In the wild, trees, shrubs, perennial plants and annuals all depend upon seed for their continued existence, and it's fascinating to discover the plethora of different methods plants have evolved to disperse their seed. Useful too; recognizing how and when seed is spread is crucial to ensuring a successful harvest, because observing what happens naturally is the first step in understanding what our best practice should be. Some plants use parachutes, others rely on propulsion and there are even plants that employ third parties to spread their seed, by attaching themselves to fur and feather with hooks or providing a delicious coating in the form of fruit or berries to attract and reward distributors.

When the seeds of the opium poppy (*Papaver somniferum*) ripen, the apertures at the top of each round seed pod open, and with every sharp knock the tiny, round seeds are forcefully propelled up and out into the surrounding area. The seeds of most poppies and many crucifers, including brassicas, wallflowers and sweet rocket, are spherical – all the better to roll across the ground away from their parent and lodge themselves in a propitious corner.

Plants like the horned poppy (*Glaucium flavum*) and the Californian poppy (*Eschscholzia californicum*) have pods that are linear and long – in the horned poppy they may be up to 30cm (12in). When it is ripe, the pod splits in two and rolls back on itself, flinging the seeds far and wide. *Lathyrus vernus* is even more fun. In a good year, there may be hundreds of delightful pea flowers on an established plant. In the species they are purple, in 'Cyaneus' more blue, and in the prettiest form, 'Alboroseus', pale pink. Most of the flowers will set seed, and the polished mahogany seedpods are an extra attraction. If you want to make more plants, though, you must forego at least part of the display and cut off a number of pods when they are in their prime, stowing them away in a paper bag or box. Leave them for a week and you will have missed the boat: the pods split in two parts that twist around in a double helix, propelling the seeds away as they do. Lupins exhibit the same behaviour. Just as with peas and beans, providing the seeds have reached maturity inside the pod, it will not affect their viability to collect seed when the pods are slightly green.

Everyone knows how dandelions spread their seeds, and what a clever

DIVERSE METHODS
OPPOSITE *Geranium eriostemon* **pods catapult the seeds far and wide. Sometimes you can even hear the explosions. BELOW Dandelion clocks are wafted away by the wind to pastures new.**

LATHYRUS PLANTS
One of the prettiest of spring performers, *Lathyrus vernus* 'Albo-roseus' makes short, bushy plants that are smothered in flowers, an excellent source of nectar for early flying insects. By early summer, the pods have ripened. The loaded stems can be snipped at an opportune moment and stored, intact, in paper bags.

mechanism it is! Each seed is provided with a gossamer-light parachute, which is jettisoned as soon as it makes contact with a stem or twig, and a hook to anchor it firmly to the ground when it makes landfall. It is mainly daisies (members of the *Asteraceae*), that use this method. Dandelions, corn-cockle, and salsify all form clocks, and seeds drift away a few at a time as they become ripe. Another clock-maker is *Erigeron karvinskianus*, the opportunistic little Mexican daisy that sets up colonies on vertical walls, finding a foothold even on the sheerest drop. Often its parachutes are caught in spiders' webs. I am frequently asked to spare a plant of this, and although it is so prolific, I always pass on a few seedheads rather than pull out a plant.

In many asters, seed ripens within the calyx, which opens wide on a bright, sunny day to allow the seed to leave in parties, until all have flown the nest and its silver satin interior is revealed. *Aster corymbosus* has particularly attractive seedheads.

Some plants employ less energy to disperse their seed. Members of the scabious family have no seed capsule: when seeds are ripe and dry, each one separates and falls from the seedhead, drifting on the breeze. The spherical seedheads of the scabious are a masterpiece of efficient design. The uppermost seeds ripen first and, if left to their own devices, will detach at the point when they become sufficiently papery and light to float off, hopefully to some new piece of ground far enough away for them to establish and flourish without competing with the parent plant. Subsequently, each tier of seeds peels off and floats away in turn. *Scabiosa ochroleuca* and closely related *Knautia macedonica* flower for months,

LATHYRUS SEEDS
As the pods dry, they twist and split, exposing round seeds, speckled like gull eggs. The seeds can be separated by hand in a tray and stored or sown. If they are to be sown immediately, soak them in an eggcup first to soften their coats.

so their seed can be collected over a long period. The first flowers tend to be the strongest and produce the most viable seed, but later flowers will also make plenty of good seed. The seedheads of both are attractive – they need not be dead-headed – and make nutritious food for small birds, especially finches.

It's sometimes hard to tell when the flowers of grasses have become seed. In most cases the colour of the stems and inflorescences change to golden brown and, since seed is heavier than flowers, the stems bend outwards.

Delicious fruits in tempting colours attract distributors when seeds are ripe. Most package their seeds safely inside this, but anyone who has studied a strawberry, before popping it into their mouth, will know that their seed is attached to the outside; herbaceous potentillas are the same, though on a much tinier scale.

Collecting seeds

We always associate autumn with fruitfulness, the time at which seed is set, and to a large extent this is true.

HANDS ON
TOP LEFT Strip the seed of *Anemanthele lessoniana* and other grasses from stems using your nails.
TOP RIGHT Grab fluffy seeds, like this *Pulsatilla vulgaris*, just as they are about to fly away.
BOTTOM LEFT Seed from members of the *Apiacieae* can be collected and sown when green, as long as it is fully ripe.
BOTTOM RIGHT Collect papery seedheads, such as those of *Aconitum* 'Ivorine', individually, or snip and store the whole stem.
OPPOSITE Sever the stems of umbellifers like *Bupleurum longifolium* as the bracts begin to dry.

There is a flurry of excitement and activity in autumn as leaves start to change colour and seed ripens. But collection can be carried out throughout several months of the year. Seed of flowers that bloom in the spring, (tra-la) hellebores or acquilegias, will be ready months before eryngiums or monkshood.

If we are taking our lesson from nature, we need to move in as soon as seed is fully ripe and about to disperse, but before it has a chance to do so.

The most crucial point about collecting seed is to gather it at exactly the right time: with very few exceptions this means when it is ripe. Seed is designed to be dispersed naturally when the conditions are best for its survival, it is also at this moment that it is best collected. The moment at which the parachutes that form a dandelion clock are wafted away by the wind is the optimum opportunity for that seed to be dispersed. As the spherical heads of opium poppies ripen they turn from

green to brown, and the seeds inside begin to rattle at the slightest touch. Remove them from the plant before they are ripe and they may never germinate; on the other hand, if we leave it too long we may miss them completely. There are no hard and fast rules; close and regular observation is the best policy.

With the majority of plants there is plenty of leeway, and given a bit of luck (a much-underestimated factor in gardening) there will be more seed than we can possibly use. But in a few special cases there is a very short optimum period to collect certain seed. If I do not move in quick-sharp on the fat seed pods of *Trillium chloropetalum* (dark and mysterious and one of my favourite plants in the garden), a whole year is lost – and when it may take as long as seven years to see a

flower on your seed-grown plants, it is almost worth camping out to beat the slugs to it.

The other factor that impinges on seed collection is the weather. Seed must be collected on a dry day, and at a time of day when the seedheads themselves have no trace of moisture or dew on them.

The equipment needed is simple: sharp secateurs, or a good strong pair of scissors, are the best tools and a large supply of old envelopes or recycled paper bags is essential. Make sure that there are no holes or open seams through which precious seeds can escape. Always try and write on the bag or envelope before collecting the seed – it's easier when it's flat!

In some cases individual seedheads can be carefully removed, but in others

VARIED FORMS
Seedheads come in innumerable forms. Sometimes seed is encased in papery pods or capsules, sometimes it is massed around the stem without any receptacle.

it's better to go for the all-or-nothing approach. You can sneak up on poppies as their seedheads start to rattle and capture several stems in one fell swoop by placing a paper bag over the lot, tightly gripping the stems together, snipping or cutting them and swiftly turning the bag right way up again. Other fine seed in capsules, such as aquilegias or *Tellima grandiflora*, is also best collected in this way, placing a bag over the seedhead when it first begins to rattle. Most umbels, members of the *Apiaceae* family, bear separate seeds at the end of individual stems. They ripen at the same rate, and again the whole seedhead can be cut with a length of stem attached.

Store any seed you gather with the stems still attached in a spacious paper bag and hang it upside down in a dry airy place so that the ripening process will continue.

The great majority of the buttercup family (the *Ranunculaceae*) make big, waxy seedpods that are generally clustered together in a head of several pods. Hellebores are an excellent example of this, and it is imperative that you check plants daily to ensure that you are there to collect the seeds at the very moment the pods begin to burst, especially in cases where you have hybridized your own hellebores. On several occasions I've had to don my spectacles to try and find the black shiny seed that has escaped – it's a thankless task. A few members of *Ranunculaceae* buck the trend: clematis and pulsatilla both rely on the wind to take their seed to pastures new, and each seed has its own fluffy tail to carry it on its way.

A close eye needs to be kept on other seeds that have the tendency to run

away as they ripen. Campanulas have an esoteric characteristic, dispersing their seed through small holes in the back of the seed pod. Make sure to catch them in time. Geraniums are escapologists *par excellence*, catapulting their seeds yards away from home to start their new lives. Euphorbias have the same ability, and in both cases you may hear the seed pods popping on a hot sunny day – move in quickly if you do. The disconcerting thing about euphorbia seed is that, should you decide to pre-empt the process by trying to take their seed early, the pods will not yield their cargo even when hit with a hammer! In this case, a paper bag tied tightly over the whole head as soon as the seeds begin to pop is the most efficient method of collection.

**RIPENING SEEDS
TOP LEFT** The winged heads of *Clematis recta* 'Purpurea'.
TOP RIGHT Gathering seed of *Hemerocallis lilioasphodelus*.
BOTTOM LEFT Catch the fine seed of *Semiaquilegia ecalcarata* in a bag.
BOTTOM RIGHT The first ripe seeds have already taken flight from this *Scabiosa ochroleuca*.

Cleaning and storing

Try to clean gathered seed promptly, especially if it is to be stored for some time before sowing.

Old seed cases and capsules sometimes retain moisture, especially if they are a bit green, and if seed pods start to rot it can spoil the seeds. As soon as possible, the seed should be separated from the capsules, or the surplus material that surrounds them. Most dry seed is fairly easy to clean. Geranium seed, for example, will usually separate itself from any remaining husk spontaneously, but if not it is easy to clean off – finger nails are a boon.

Larger seed, especially anything leguminous – lupins and sweet peas are prime examples – will often fall out of the seedpods as the latter dry out. Labiates, such as salvias, usually make up to four large seeds inside each calyx. If these don't fall out by themselves, gently peel back the calyx and just push off the seed.

Stems of most umbels can be stored hung upside down inside a paper bag or a large envelope to ripen fully before cleaning. Fennel (*Foeniculum*) is a typical example, or *Orlaya grandiflora*. In the case of the latter, each seed is protected by a barbed seedcoat. If you have collected these or any fine seed in capsules by placing a paper bag over the entire seedhead, when it comes to cleaning them the contents of the bag should be spread out on a sheet of paper. Newspaper will do, but it is easier to see what you are doing on a piece of plain paper. Fold the paper

down the centre before you start, to allow easier decanting into packets when the seed is clean. Detritus can be picked away by hand, winnowed by gently blowing and/or placed in a sieve and rubbed gently to dislodge any dross.

Winnowing is an ancient practice to separate seed from chaff. Although it is done on a much tinier scale for garden seeds than it would be for harvested grain, the principle is the same. Sometimes seed can be taken outside on a breezy day to start the process, but this can be fraught, especially when a surprise gust of wind fails to discriminate between seed and chaff! Some seed, such as campanula and lobelia is very fine, and it may be difficult to remove the last vestiges of powdery dust without losing the seed itself. As long as everything is dry, it should still store well even with a bit of the chaff left in.

SIFTED AND STORED
Blow lightly to winnow, separating seed from detritus. Seed should be clean before it is packed away. Until then, it can be kept in paper bags, hung up in a dry, airy place.

ARMOURED SEED
European sea hollies, like this *Eryngium bourgatii,* protect their seed from grazing animals with dramatic, spiky, prickly bracts. Cleaning them can be a painful business, so we use the blunt end of a pencil to push the seeds out.

The European sea-hollies, or eryngiums, have an ingenious construction, with a cone of tiny flowers sitting on top of a circle of bracts. This is almost invariably spiny and protects the flowers and seeds from grazing animals. It works on human beings too, and collecting and cleaning eryngium seeds is one of the least popular activities at Glebe Cottage Plants. Seedheads are cut with a length of stem, just as the seed begins to detach itself from the central cone. The easiest way to persuade the seeds to fall away is to push them off with the blunt end of a biro or a pencil. It is still a tricky business, but the prospect of all those new plants of *Eryngium bourgatii* (blue form) with its deeply-cut, marbled leaves and vivid caerulean-blue bracts is more than enough to make it worthwhile. Stems can be laid in wooden trays or cardboard boxes for a few days to encourage the seed to loosen itself.

Closely related to eryngiums, astrantias are a far less painful proposition. Their dancing pincushion flowers turn almost imperceptibly to oval seed and if they are left, eventually fall from their hair-fine stems and collect inside the papery bracts. The best time to harvest astrantia seed is when it is still predominantly pale green, but tinged with brown and just beginning to fall. Stalks comprised of several branching stems can be cut when the majority are ripe. If you have a particularly precious plant, each seedhead can be harvested individually as it ripens in succession and can be stored in the same bag.

You can start sowing cleaned seed immediately if you have the facilities. Otherwise, store seed in envelopes, writing clearly the name of the variety and the date of collection and cleaning; it is worthwhile recording these on the label you eventually use when sowing the seed. Recycle envelopes and old paper bags to store your seed, and never keep them in plastic bags, where they would sweat. Small cardboard boxes are efficient for large seedheads – shoeboxes are perfect – but don't close the lid until the seed is properly dry.

Store packets in a cardboard box or an airtight tin or hang them up in bags in a cool dry place with a buoyant atmosphere, out of direct sunlight. Uneven temperature during storage is one of the main reasons seed fails to germinate when planted out. At first sight, the greenhouse might seem ideal, dry and airy, but temperatures in glass structures fluctuate hugely, and the atmosphere may be humid from time to time. A definite no-no.

Enlightened sections of horticultural societies now recognize how vital saving seed has become. Projects, such as the Kew's Millennium Seed Bank, acknowledge the importance of ensuring biodiversity and conserving our heritage by storing as many varieties of seed from around the world as possible.

A GOOD HAUL
Buckets and boxes of envelopes and bags of seed from the garden are the sign of a successful seed-saving afternoon.

Agapanthus

How exciting to grow the exotic

Agapanthus is easily grown from seed. New plants may flower in the second year after sowing and most of the offspring will be similar to the mother plant, although if you want truly identical plants they will have to be vegetatively propagated by division. Other exciting bulbous and semi-bulbous plants can be grown from seed; eucomis and dierama, both South Africans, like the agapanthus, are two of my favourites. Agapanthus seeds are most intriguing – black with tails, almost like tadpoles.

1

2

3

4

5

'Lily of the Nile' from seed, and so easy!

6

7

8

9

10

SEED TO FLOWER

1 Pick the seed just as the first seedpods open.

2 Agapanthus seed is black when fully ripe.

3 Seed can be picked out of the seedheads and sown immediately.

4 When sown, cover the seed with grit.

5 As seed germinates, the old seed coats can be seen.

6 Gently pull the contents of the seed tray free.

7 Each seedling has strong roots and is ready to be potted up.

8 Ease each seedling into its own pot.

9 A tray of strong young seedlings.

10 Agapanthus and eryngiums both from seed, ready for planting out in the garden.

Overcoming dormancy

The dormancy of seeds is a very complex subject and tackling it in depth is outside the scope of this book.

But there are fairly standard, simple ways of 'breaking' dormancy. They are relevant to growing all sorts of seeds, but particularly trees and shrubs. Seeing any seed germinate is a thrill. Finding the right place for it when it is capable of going it alone is equally rewarding, but if the plant you have reared becomes a permanent feature in the garden you may go on to develop a lifelong friendship. Growing trees from seed you have collected is a wondrous experience, whether it be a single specimen or a stretch of native hedge. Raising trees from seed is rewarding and although there may be no room in most gardens for forest or woodland species, native hedges comprising a wide range of indigenous trees and shrubs are becoming a feature in more and more gardens, not just in rural areas but in the heart of cities and suburban areas too.

It may be gratifying, but it is not necessarily quick or foolproof. Dormancy is a feature of most tree seeds, evolved to ensure that seed does not germinate until conditions are right for successful growth and development. In some cases seed has a hard coat; many leguminous trees and shrubs fall into this category, including laburnum and cytisus. In nature, the actions of fungi and bacteria break down the seedcoat, let water in, and allow germination. Other seeds do not develop beyond a certain stage, remaining immature until warm temperatures kick-start development

and enable the seed to germinate. Many seeds need a period of cold to start a series of chemical events that, with the right temperature, enable seed to germinate. Seed may need just one of these conditions, or two or three. There are simple steps, though, to help berries, nuts and assorted tree seeds to germinate more readily.

SCARIFICATION

This is simply a process to abrade the hard seedcoat and either break through it or thin it down so that the seed can take up water. Larger seeds that can be handled individually can be nicked with a penknife to break through the seedcoat. Coarse sandpaper or a file can be used too; it is probably easiest to keep the sandpaper flat and rub the seed across its surface, though this sometimes results in an accidental manicure. A safer method is to line a jam jar with coarse sandpaper, drop in the seeds, close the lid and shake vigorously – it's good exercise. As long as the seedcoat is partially worn down, this should allow water to be absorbed.

STRATIFICATION

Stratification sounds very technical, but as with all aids to propagation it emulates a natural process. Nature allows for a high degree of failure: if every hig on every hawthorn tree grew into a new tree we would be overrun, but very few become trees in their own right. When we grow trees from seed we can offer each one customized care.

WINTER CHILL
Even when the seeds have flown, the architectural forms of many seedheads remain. The snow that crowns these allium skeletons will also be chilling any seeds in the ground.

The embryo in every seed contains an inhibitor that prevents the seed germinating until the time is right. This is usually when the weather starts to warm up after a prolonged period of cold; in other words after winter and into spring. It is a chemical change that takes place, rather than a physical one. By replicating the process in a controlled environment, we can not only maximize the seed's chance of success, but also telescope the whole process into a shorter time.

The easiest way to grow these seeds is to plant them in the open ground after collection. But in mild winters temperatures may not be cold enough for long enough to ensure the seed germinates the following spring. The answer is to provide your own winter by chilling the seed. Mix seed with four times its volume of sieved, damp leaf mould (peat was always recommended for this in the past) and put the whole lot into a polythene bag. You can add coarse grit if it seems too cleggy. Label the bag and leave it in the warm for a couple of days while the seeds absorb enough moisture to make them swell, then put the bag in the coldest part of the fridge for two months.

Sow seed in a seed bed. Each tree and shrub has its own idiosyncracies but this practice will do the trick for most of their seeds.

Sowing in containers

Nature ensures that every seed is in with a fighting chance, and plants have evolved to make distribution and germination as successful as possible.

Most plants make prolific seed, far more than they need, as an insurance policy. Seeds will have to contend with so many trials and tribulations in the wild – from drought and flood, to hungry predators and stony ground – that only a few will be successful. Even if they germinate, their future is fraught. Although success cannot be completely guaranteed, when we grow from seed we can give them the conditions most conducive to success and oversee their development, nurturing them and seeing to their every need until they are big and strong enough to fend for themselves.

CONTAINERS AND COMPOST

The most foolproof method is to sow in seed trays or pots. We use the same seed trays year after year, washing them out and drying them. Mostly they are half seed trays, made of rigid, durable black plastic. They are strong, stable, and manageable. It is tempting to use full-size seed trays, but they are simply too big. Half trays have a big enough surface area to give a good number of seedlings a sound start and plenty of elbow room, and are deep enough to allow development of a strong root system. Anything can be pressed into service, of course – margarine tubs, punnets, and so on – and I am sometimes accused of not fully embracing recycling. But since many of our half trays have been used upwards

of 20 or 30 times, I don't lose too much sleep worrying about it. They have played their part in the genesis of a lot of new plants and hopefully they will outlive me and someone else will use them.

About 20 years ago, when the trays were new, Neil made me a presser board. It is a simple, but very effective tool, a rectangle of plywood with a piece of dowel (I think it was a broken broom handle) sliced along its length and attached to the plywood. The board's corners are slightly rounded and it fits the trays exactly. It is used to gently tamp down the compost and firm in seed, usually with a layer of grit over the top. Fill your seed tray or pot to the brim then gently press it down so that the compost is just below the top of the pot (if you are sowing into pots, another slightly larger pot will do the job well).

Which compost to use? Opinions vary, so it's worth experimenting. Compost for seed sowing must be fine, so seeds are not accidentally buried – we are sowing them, not committing them to an early grave. It must be sterile too, so there is no competition between germinating seedlings and antipathetic organisms, including weed seeds. When I first started to grow from seed I experimented with sterilizing my own compost in the AGA. It was an anti-social exercise: the smell was not altogether unpleasant, but decidedly odd and unlike food, the ovens were monopolized for long periods of time,

RECYCLED MODULES
The insides of toilet rolls make efficient sowing pots for larger seeds. They are suitable for station sowing, and disintegrate in the ground. These broad beans were collected from our own plants the year before. They can be planted out a month after sowing.

and inevitably the receptacles that were pressed into service were baking trays or cake tins.

Seed compost has to allow seed to germinate, maintain moisture without ever becoming waterlogged, contain plenty of air and have the wherewithal to feed the seedlings from when they emerge through to the point where they will be separated and pricked out. Some gardeners prefer to use a proprietary mix, specially prepared for sowing seeds, others use a multi-purpose compost. Compost can contain mixtures of many ingredients, from bark to sand and loam. Some of the peat-free mixes do not divulge their contents – they tell us what is not in the compost, but not what is! You must let your conscience decide, but since you don't need peat why use it? It is worth remembering that a quantity of sterilized loam in your compost will keep seedlings going longer.

SOWING

Most of the seed we sow is relatively fine. Put a pinch or two into your palm, or onto a saucer, to get a clear idea of how much you will actually be sowing. Though it sounds tedious, it's a salutary lesson to count 50 seeds to appreciate just what is in a tiny pinch, even if you only do it once. If we have ample seed, all of us are tempted to use too much – just in case. But this almost

MODULE TRAYS
Module trays ensure minimal disturbance when potting on. Drop the tray onto your table to settle the compost, sow your seed, covering larger ones, and top with grit.

always results in damping off, a fungal disease that causes seedlings that are a picture of health one day to collapse and keel over the next. 'Thou shalt not sow thy seeds too thickly, or they will die!' is one gardening commandment.

Sow seeds as evenly as possible over the surface of the compost. Start with a fine sprinkle around the edge and work back and forth across the tray or pot. If seed is really fine, you can mix it with a little dry sand (dry a small quantity just for this purpose) before sowing. The sand will do no harm, and will enable you to see just where your seed has gone and give you a better chance of distributing them more evenly.

Debutantes in the seed-sowing game are often flummoxed by how to cover the seed. Many books advise covering all seed with a layer of sieved compost, perhaps 1cm (½in) deep, but for almost all the seed sown here at Glebe Cottage we do not use compost to cover it at all, but a fine, even sprinkling of dry grit. This replicates the conditions the seed would experience naturally, allowing light and warmth to get to work on it, and provides emerging seedlings with sharp drainage, while retaining moisture under the surface where it is most needed by developing roots. In all aspects of gardening the best advice is to emulate nature: this is especially true for growing from seed.

But what about big seeds? In nature big seeds also fall to the ground and, although many of them are round (think of peas and sweet peas) and roll off to lodge themselves in nooks and crannies, they do not bury themselves. Once they send out their first root (called a radicle), they are pulled down into the soil. We can make that happen by gently pushing the seed into the

compost so it is covered by about the same depth of compost. Many of these larger seeds, like lupins, sunflowers, beans and peas, are attractive to mice, so hiding them is a partial deterrent to full-scale theft. Instead of random scattering in a tray or pot, most big seeds lend themselves to station sowing. This means sowing each seed individually, and the best equipment for this is a module or cell tray. They come in all shapes and sizes, or you can make your own by compartmentalizing a seed tray with partitions of plastic, wood or cardboard. Station sowing individually means that developing roots have no competition from their peers, and it has advantages when it comes to potting on. When they are ready, seedlings can be lifted without any disturbance to their roots. If you are using a module tray, fill it to the brim, strike it off with the side of your hand or a piece of wood and drop it sharply onto a bench or table to settle the compost. Push in the seeds, drop the tray once more to cover them with compost and grit the surface.

Watering from overhead can be disastrous, washing seeds to the side of the container and undoing all the careful work of sowing them evenly. Watering evenly and thoroughly without disturbing seed is a difficult feat to pull off with a watering can, no matter how fine the rose. The best idea is to stand trays and pots in a bowl of shallow water to let them imbibe gradually. When the grit becomes damp, they can be taken out to stand and drain. Not only are the trays or pots watered thoroughly, but as the water drains away, the seeds are pulled down into intimate contact with the compost.

Sowing green seed

As a rule, in a temperate climate, seed germinates in spring – but not always.

Most seed will have ripened and been dispersed during the autumn and lain dormant to be awakened by spring rain and increasing temperatures. We usually collect seed when it is ripe and dry, but there are a few exceptions to this rule. There are some plants whose seed can be taken and sown while it is still green, which is counter-intuitive as we are not emulating nature.

Members of the primrose clan, ranging from the exotic candelabra primulas from Japan, China and the Himalayas to our humble native primrose (*Primula vulgaris*), all respond well to this treatment. In the normal course of events, as primrose flowers fade and petals disintegrate, flower stems lengthen and weaken. With the added weight of the seeds it eventually lies along the ground. The seed capsule swells and during summer becomes brown and dry. During autumn it disintegrates, and the brown seed lies on the earth. Over winter the cold breaks the seed's dormancy, and by spring it starts to germinate. Gardeners can steal a march on nature by collecting seed when it is properly formed, but still green, before dormancy ever sets in. If fresh seed

SUMMER SOWINGS
Take swollen, but still green, seed pods from primulas and gently puncture the fat capsules. Squeeze the green seed onto the compost surface and cover it with grit. Germination is rapid, and soon the seedlings will be big enough to prick out.

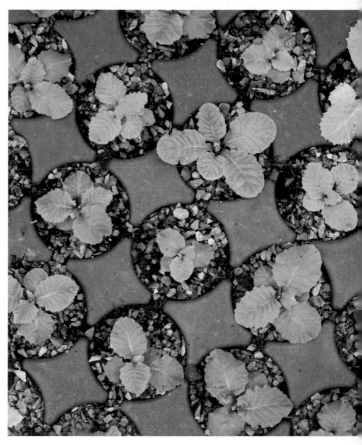

MOVING UP
When young plants have well-developed root systems they can be potted on.

PERMANENT HOME
Planting out is the best part of the whole process. To see a bank of primroses that you helped into life is truly rewarding.

from any of the primrose family is sown straight away, it should germinate within a few weeks.

Take a few seed pods from primroses when they are fat, but still green, tear back the outside casing of the old calyx, then burst the membrane with your nail and squeeze gently to expose the seeds. Carefully remove the seed – it may be sticky, in which case you almost need to scrape it off with your nail – and sow it on the surface of damp compost. Cover it with a thin

layer of grit and, after watering, leave it outside in a shady site where the rain cannot wash away the seed.

Germination is rapid, and after a few weeks you should notice a green haze on the surface of the tray as the seedlings begin to emerge. They will grow quickly and can be pricked out and potted on as soon as the true leaves are visible. By this stage each seedling will have its own root system. Module trays are best, giving each little seedling the opportunity to grow on and expand. If plants are big enough, they can be potted on into individual pots during late summer and planted out in autumn, but I prefer to leave them in their pots over winter and plant out the following spring.

This is a rapid way to build up a colony of primroses, or a mass of candelabra primulas beside a pond, or a wealth of cowslips in a chalky meadow planting. It is a wonderful feeling to create a lavish display with plants you have grown yourself.

HAPPY PLANTS
Just a few months after first being pricked out, each plant has developed a strong crown and is beginning to flower.

Pulsatilla

Many alpine plants work their roots deep into

The pasque flower (*Pulsatilla vulgaris*) is now rare in the wild, where it makes a neat plant with furry blooms on silken leaves. It is a familiar alpine in gardens, though there an easy life can make it leggy. Gritty, limey soil and a sunny, dry, exposed site ensure flowers – and seeds. The fluffy seedheads are just as tactile as the flowers, and they are the best way to increase numbers, because pulsatilla resents being disturbed once it has got its roots down. Seedlings will be ready to flower a year after sowing.

1

2

3

4

5

crevices and cannot be divided – seed is the best way to increase them.

6

7

8

SEED TO FLOWER
1 The silky seedheads of *Pulsatilla vulgaris.*

2 Pull the seeds away from the stem as they begin to fluff up.

3 You can just pinch the entire seedhead between fingers and thumb.

4 Errant seeds sometimes have ideas of their own.

5 Push the whole seedhead into a clay pot filled with seed compost.

6 A layer of grit keeps compost moist and weighs down the seed.

7 Looking after mature plants ensures good flowering and further seed.

8 Who wouldn't want even more of the delightful pasque flower?

Pricking out

Once seed germinates, keep a keen eye on seedlings. Almost overnight, it seems, they outgrow their space.

As with all propagation, sowing your seed is just one part of a larger process – all the stages are interdependent and equally important in its continuing success. When the seedlings have developed their first true leaves, they should be pricked out promptly. Hopefully they will be a picture of health at this stage, but they are growing rapidly and can soon deteriorate. It is crucial that they are separated and given their own 'room' as soon as possible.

In the case of dicots the first leaves that develop are the 'seed' leaves or cotyledon leaves. In most cases these are a pair of simple, generally rounded leaves that look nothing like the true leaves of the plant. Many of them have evolved to photosynthesize, starting the process of the new plant feeding independently rather than relying on food resources within the seed. In contrast, monocots, such as grasses, irises and many bulbs, do not produce seed leaves, but send up a typical linear leaf. This comes not directly from the seed, but from the meristem – the cluster of rudimentary first cells that will eventually produce all the growth the plant makes above the ground. Some say that perennial monocots should not be pricked out until their second year. We sow monocots with big seeds, such as dierama, hemerocallis, eucomis and iris, direct into modules or into seed trays, but prick them out into their own separate compartments before their roots have become entangled.

Most of our seeds are from dicots, however, sown in seed trays and easily recognizable by their twin seed leaves.

SEPARATING SEEDLINGS
Once separated, these seedlings of *Geranium wallichianum* 'Buxton's Blue' are potted into home-made modules with cardboard dividers in a seed tray.

LIFTING SEEDLINGS
Eryngiums germinate very successfully. As soon as the seedlings have true leaves they can be eased from the tray and pricked out. Lift out one small clump at a time.

After these, the first true leaves begin to appear. When the great majority of seedlings in the tray have developed one or two, bang the seed tray on the bench to loosen the compost and to separate it from the tray. Take a chunk of seedlings, compost and all, and separate each one using the seed leaves to gently tease them apart. Once a seedling has developed true leaves, cotyledon leaves are expendable so they are occasionally sacrificed in the rigours of separation to ensure that stem, roots and true leaves are undamaged. On most occasions they remain undamaged and their life continues, though in most cases, once the seedling settles down, they will gradually fizzle away of their own accord, having fulfilled their role. The seedling is gently lowered into its new home. Very young, well-spaced seedlings can be gently lifted from the compost by using a stick or dibber (a chopstick works well).

Traditionally, seedlings would be pricked out into a large seed tray or box, but pricking out each seedling into an individual compartment of a cell-tray instead will allow it to grow unhindered. After a few weeks, they should have filled their module compartments and can be moved into more spacious accommodation without root disturbance. In a box or seed tray, there is yet another disruption when seedlings have grown on. They must be pulled apart and separated, and although the great majority survive, this upheaval can set them back.

Seedlings should be pricked out into compost one degree stronger than that used to sow the seeds. So if you were using a John Innes mix, you would have used John Innes seed compost for your initial sowing and John Innes No. 1 for pricking out. There would be a little more fertilizer in the No. 1 mix that will help the fast-growing young seedlings to develop. An alternative to

this rule is to use the same compost as that in which the seed was sown, watering in well and subsequently applying a very dilute liquid feed.

Despite the best of intentions, sometimes young plants get far too large in the original seed tray or pot in which they were sown, but with extra care they can still be carefully teased apart and planted directly into small pots. Occasionally we do this deliberately with seed sown in the autumn. Eryngiums, codonopsis and platycodon will germinate well from an autumn sowing, but they may not survive being disturbed and separated just before the winter, when they take a rest and go into gentle hibernation. When new spring growth appears, they can be pulled apart and potted up. The roots may be slightly more tangled than they would have been in the autumn, but the plants will be tougher since they are a little older and better able to recover from the disruption.

SETTLING IN
Gently tease out individual seedlings. Each young plant has a strong root system and a real lust for life at this stage.

Within weeks after pricking out they will already have filled their alloted space in these modules and be ready to move on again.

Sowing outdoors

Before the advent of greenhouses – or at least their wide use by amateur gardeners – most seed would have been sown outside, in the ground.

Although pots have been in use for centuries, most of the seed sown in the past would have been sown direct, straight into the ground. We still sow most of our vegetables direct and sowing biennials outside is a trusted and true method of raising such spring stalwarts as wallflowers (*Cheiranthus*) or sweet Williams (*Dianthus barbatus*). Many other plants can be raised like this, from foxgloves (*Digitalis*) to stocks (*Matthiola*), and there is no reason why perennials cannot also be raised this way.

There can be a few problems. On the minus side of the equation, plants will have to fend for themselves to a much greater degree than those in trays and pots. Germination may be sporadic and uneven and seedlings will be prone to the ravages of slugs and caterpillars. They may be overcrowded or have to compete. On the plus side, because they have their roots in the soil there is less chance that short-term neglect will have such immediately disastrous consequences as it might for seedlings in a tray or pot.

SOWING IN ROWS

There may be no need to create a seed bed before you sow. If you are already growing veg, the job has probably been done. In any case, preparatory work

BIENNIAL SEEDBED
Mark out a shallow drill with a trowel, using a taut line to keep it straight.

SOWING SEED
Sow thinly, a pinch at a time, to allow seedlings adequate space to develop.

COVERING UP
This little rake is invaluable for running down the drill after sowing, returning just enough soil to cover the seed.

AUTUMN SOWING
**In the early days of
autumn, seeds of
hardy annuals can
be broadcast sown
onto any empty
spaces in the border.**

is straightforward; remove weeds, especially perennial weeds, and rake the whole area to loosen top soil and create a flat, even seedbed. Tamp the area down with the back of the rake so there are no air-pockets. Try not to tread on it, but work from a plank, especially if your soil is heavy and wet.

Draw a shallow drill using a stick along a straight edge. Some people water the seed drill before sowing, arguing that it is the seeds they are sowing that have the benefit, not weed seed that may be present in the surrounding area. Once seed is sown, soil can be gently returned to the row and raked along its length. Always rake along the row, so even if you move seed accidentally it will still be in the line. It's quite easy to distinguish your seedlings from weeds, especially since the former will be growing in straight lines, and to pull out the offenders so

they do not deprive your new seedlings of water, light and nutrients. If you sow thinly, you should not need to thin your seedlings. When they are about 5cm (2in) high, lift and transplant them with plenty of space between, both in the row and between the rows. It is probably best to lift with a small hand fork and dig individual holes with a trowel. Transplanting is traumatic, so always water thoroughly; your little plants will need a drink to settle them in.

Seed for biennials is traditionally sown in late spring. It germinates rapidly and, with one intermediate transplanting, plants should be large and strong enough to be lifted and transplanted to their permanent quarters. When the gardeners from big estates were growing their wallflowers for the next year's big spring bedding display, they would lift and replant seedlings several times and trim back the roots, sometimes replanting onto beds of slate buried not far below the surface of the soil. This produced a fibrous, branching root system and resulted in bushy plants loaded with flowers. A painstaking business, but labour was cheap and impressing visitors was all-important!

BROADCASTING SEED

Throwing seed around in great armfuls is a liberating experience and, although we may not get the opportunity to follow in our ancestors footsteps by scattering seed across huge swathes of ploughed land, there are sometimes opportunities, even in a small garden, to indulge in this empowering experience. Most frequently this method is used to sow annuals – flowers that go from life to death during just one year. Traditionally, annual borders of this ilk were an economical way to create a kaleidoscope of colour. It was quick, easy and cost only the price of a few packets of seed. This is still a fairly common practice, though nowadays

it is far more likely that we would be sowing an annual meadow, with all our cornfield annuals mixed together and broadcast over an entire area, than sowing separate varieties in their designated patches as part of a grand design. An annual cornfield mix might contain field poppies (*Papaver rhoeas*), cornflowers (*Centaurea cyanus*), corn marigolds (*Glebionis segetum*), corncockle (*Agrostemma githago*) and, perhaps, the corn itself. In a traditional annual border, areas can be outlined in sand, each containing just one variety. It is lovely to experiment with this idea, but in a more modern spirit, allowing seed to mix and mingle at the boundaries creates a far softer picture than the crisp edges of a scheme where seed is not allowed to stray away from its allotted area.

Where edges are encouraged to overlap, seed of each variety should be thinned out so there is a neighbourly mixing, rather than a battle for space. In a scheme like this there will be little opportunity to move in later and thin

A SEEDBED FOR TREES AND SHRUBS

Because tree and shrub seed takes longer to germinate than that of perennials, it is best to sow it outside. Seed compost in pots and seed trays may become stale, and seed in the ground is subject to the activity of fungi that ensure healthy tree growth, forming complex symbiotic relationships with the developing roots. This process is promoted by including leaf mould

when preparing the seedbed. A seedbed for baby trees needs very little attention and you can pack in large numbers without any deterioration, provided the seedlings are transplanted during their next dormant period. You could raise a whole oak wood in a coldframe!

The seedbed can just be an area of the vegetable garden or a flower bed. If you build

a simple box then the soil can be prepared to make sure it is well aerated (seedlings need oxygen). This should also ensure better drainage – seedlings will die in waterlogged soil. It is easier to keep seed safe from the depredations of mice and voles too.

Another advantage of a seedbed with edges is that you can construct a simple rigid

frame on the ground, covered with clear plastic or, if you are lucky enough to have them, Dutch lights or old windows. Temperature is an important factor in the chemical process of seed germination. Warmer temperatures usually result in speedier germination.

Most tree seed is big, so station sowing is easy. Position on the surface then press

them into the surface of the soil. A plank or your seed presser board is ideal. A fine layer of grit over the surface, just as in the case of seeds sown in a seed tray, will ensure good drainage around new shoots, retain moisture and keep the development of mosses, liverworts and weeds to a minimum. It also makes it easier to weed without disturbing developing seedlings.

the seedlings, so it's much better to sow thinly. When seed has been sparsely sprinkled, even if there are a few gaps initially, individual plants will relish the opportunity to spread themselves.

Broadcasting is also the favoured method for sowing a traditional hay-meadow planting. In contrast to the temporary nature of an annual corn meadow, a hay-meadow is a mixture of flowering perennials and grasses. More and more of us are turning areas of our gardens into mini-meadows. Not only is it a delight to see wild flowers outside your back door, but since gardens cover a larger area than national parks and nature reserves put together they are crucial in the preservation of our native flora and the insects and wildlife that depend on them.

Very few gardeners have really experimented with using this method for a 'prairie planting', where strong grassland species from temperate

habitats grow in mutually supportive communities in a naturalistic manner. There seems no good reason why seed from perennial plants, which would grow together well, could not be combined by broadcast sowing. Again, the sowing should be sparse, but as with any perennial sowing, plants could be thinned out and moved around during the winter.

The method in all these cases is the same: seed should be roughly measured for the size of the patch and broadcast over the allotted area. To help judge where you have sown, mixing it with dry sand helps. Hold the seed in the palm of one hand and take liberal pinches with the other, keeping this hand a couple of feet above the soil surface – any closer and seed tends to end up in lines. Seed can be gently raked into the soil using hands or a rake and watered in with a fine rose on the watering can.

MIXED BORDER
The traditional way to create a border of hardy annuals is to mark out areas with a bottle full of dry sand. Label each area after sowing and water in using a fine rose.

Self-sown seedlings

There are some plants at the garden party who were never asked.

Almost all these uninvited plants are brought in by seed. Birds may have dropped it, or it may have drifted in by parachute from some far away place or simply catapulted itself from plants elsewhere in the garden or next door. Many are weeds, which we spend hours removing to ensure they don't spill their progeny into our flower beds, but if a weed can sow itself so can a desirable plant.

Not all garden gatecrashers are unwelcome guests. Some are welcomed with open arms into any venue, from a tiny urban plot to a vast rural estate. By definition, they are self-seeders. Many are annuals or biennials. Others are perennials, brought by seed wiped from a blackbird's beak, caught in fur or feather, or jettisoned from a bed yards away and now appearing magically in the crook of a wall or the space between flags. However they arrive, nobody asks to see their invitation. Purists can easily remove them should they have the temerity to interfere with grand designs. The rest of us think ourselves lucky and have the good sense to let them enjoy themselves, making the most of the gatecrashers' spontaneity.

Some of the best plantings in our garden have nothing to do with us. They made themselves! A cranesbill Geranium nestled between paving stones with a backdrop of lady ferns, a host of *Eryngium giganteum* in the gravel of a path or waving stems of

Verbena bonariensis bobbing up between tall molinias – all of these have placed themselves with no interference from us. Often these accidental combinations beat our carefully crafted schemes hands down. It would be impossible to copy the way honesty (*Lunaria*) tucks itself into corners, or borage finds a nook in which to flourish. Time and time again, when asked about a winning combination, gardeners have to admit they had nothing to do with it. 'It just put itself there', they marvel. How come these plants put themselves together in such exquisite combinations, and are always so happy?

The delight on finding self-sown seedlings for the first time stays with you for your entire gardening life. What's more you never get blasé – each time it happens, it is just as humbling and awe-inspiring. The first time I saw self-sown seedlings in my garden, I just couldn't believe it and kept going outside, just to check again that they

NATURAL HOMES
One delightful picture after another is created by self-sown seedlings. Often we leave them to their own devices, but sometimes we lift a few for elsewhere in the garden.

TUCKED AWAY
Eryngium giganteum (left and centre) and *Aquilegia vulgaris* (right) seem to know just where to put themselves: in inaccessible corners in paving.

were still there. Now, 30 years later, these self-invited plants are always treasured – after all, it's a great compliment when a plant actually chooses your garden as the place it wants to live.

Some people moan about the way in which plants self seed. Why? If there are too many they can be moved to a different site or given to friends and neighbours. The famous lady gardener Ellen Willmott gave the gate-crashing story a new twist at the turn of the 20th century. Being one of the foremost gardening pundits of her day, she was often invited to view gardens. If she thought they might be a bit boring, she would fill her pockets with the seed of *Eryngium giganteum* and sprinkle it surreptitiously here and there around the garden. The following year up would come the seedlings with large, glossy leaves – very intriguing. And the next year garden owners would be astonished as tall stems shot up from the innocuous green rosettes, topped with massive heads of spiky silver sea-holly bracts.

Anyone who has this handsome sea-holly in their garden will know that once it arrives it is there for good, and its lingering presence long after the sower had left earned it the name 'Miss Willmott's Ghost'. Although it is a biennial, dying after it has made seed, it is one of the most successful self-sowers there is. Ideal in a gravel bed or any dry area, it can grow to 1m (40in) high with branching stems, each bearing a large, silvery bracted flowerhead with a tall central cone.

Perhaps the opium poppy is the most abundant self-sower. Although its narcotic properties don't develop in our climate, it does provide us with a profusion of glorious flowers, followed by striking glaucous seedheads that can be dried successfully for winter decoration. It turns up out of the blue in all sorts of colours and forms; some are simple and single, while others have fussily ornate flowers, like a Barbara Cartland blouse – in fact one that we used to grow here was called 'Pink Chiffon'. There are reds, purples, magentas and pinks, as well as near black and pure white forms. Use them running through a bed to homogenize the planting, or just let them have their head. Sometimes they sow themselves so prolifically they have no room to develop properly, but they

can easily be thinned at an early stage – it is better to have a few magnificent specimens than a tangled mass of stunted plants.

All poppies are notorious self-seeders and have a huge ability to survive; their seed remains viable for many years, and they often germinate when the ground is disturbed after they have lain dormant for decades. *Meconopsis cambrica*, the Welsh poppy, is another member of the clan. Whereas the opium poppy and the corn poppy prefer open, sunny sites, the Welsh poppy is happiest in dappled shade. It will put itself here and there amongst hellebores and astrantias, where the sharp, citric yellow of its crumpled, tissue-paper flowers shows up brilliantly against the green backdrop. Once you have one plant it will seed itself about, or you can help it to colonize an area by shaking a couple of seed pods around when you can hear the seed rattling inside them.

If, on the other hand, your garden is sun-baked and the soil is thin and dry, the Californian poppy will revel in the conditions, which resemble its native haunts in the coastal ranges. A blaze of this poppy in vivid, sulphurous orange is a difficult picture to beat. If you are starting from scratch, just cadge a pinch of seed or buy a packet and sow it direct in spring. Poppies don't transplant well, so it's pointless to sow them under cover.

The little Mexican daisy *Erigeron karvinskianus* loves a similar situation, making colonies along walls and through pavements, and is smothered in pink and white flowers from late spring right through to Christmas. It's often encountered in the West Country and was top of my list when we first started this garden. I am frequently asked to spare a plant of this and, although it is so prolific, all we give is a pinch of seed, as the plant has a tap root and it is impossible to transplant. Sow the seed in individual modules and when well rooted push the rootballs into the crevices between the stones of a dry stone wall. Seed mixed with soil (a bit like the 'seed bombs' thrown by guerilla gardeners) and pushed into cracks and crevices works too. Ever after the erigeron will sow itself without assistance, and like all the other gatecrashers it will almost invariably end up in the right place.

ON STONY GROUND *Geranium pratense* (left), *Anemanthele lessoniana* (centre) and *Foeniculum vulgare* 'Purpureum' (right) showing that a gravel layer won't prevent self-seeding – thank goodness!

NATURAL GENIUS

Some plants have evolved extraordinary mechanisms to make and spread their seed.

CYCLAMEN

The scrolled buds of cyclamen unfurl into flowers that look surprised to be upside down. When they shrivel, the stem coils, pulling the seed pod down to the ground, where it swells then splits. The seeds have a sugary coat, and ants move in carrying off the sweet booty. They abandon it at a distance, meaning less competition for the original corm. Corms can grow as big as dinner plates, but seldom exist in isolation. With this method of distribution it is easy to see how colonies grow.

If you want more, the only way is to sow seed from a specialist or your own plants. Collect it just as the pods start to burst and before the ants get in. Sow in trays, on the surface of open seed compost with extra grit. Cover with sharp grit and stand outside or in an unheated greenhouse. Prick out when a couple of leaves are visible.

CERCIS

So many of the plants in our gardens remind us of the people who gave them to us or introduced us to them. One of my favourite trees is a mature *Cercis siquilastrum,* the Judas tree, which my mum gave to me as a young plant. She didn't travel abroad until later in her life, but when she did she was intrigued by new places and new plants. She picked up a few cercis seed pods from the pavement in Portugal, sowed them when she got home and eventually gave me a young plant. What interested us both about the tree is the way in which the vivid pink pea flowers spring from the naked twigs and branches.

PRIMROSE POLLINATION

Primrose pollination is fascinating. Any population contains thrum-eyed and pin-eyed flowers. In the pin-eyed flower (left) the female style and stigma reach to the top of the corolla tube, poking out just like a pin-head. In thrum-eyed flowers (far left), the position is reversed: the stigma is invisible, halfway down the tube, and the male anthers are clearly visible. This mechanism ensures cross-pollination and avoids inbreeding.

The big picture

Hooray – my first pulsatilla seedlings are through! The same thrill returns each time seed germinates. In the background is an *Erygnium bourgattii* with particularly purple stems – later on we will collect its seed and hope it comes true.

STEM CUTTINGS

CUTTING MATERIAL
Strong, short shoots taken from this *Lonicera periclymenum* 'Graham Thomas' will be taking root to make new plants while the honeysuckle flowers.

Sometimes you're in your garden, strolling around (or more likely dashing about), when a plant stops you in your tracks. Perhaps it's the flowers or their perfume that pulls you up short, or just the translucence of new foliage. You remember then that this is a special plant, one you have known all its life since you helped it off to a flying start from a cutting – maybe from a plant you couldn't bring with you from a previous garden, or from something you first saw and coveted in a friend's collection.

Many of our gardens are full of such plants, and somehow it is impossible not to hold special affection for them and to recollect just how and when they came into being. No matter how often you repeat the process, it never palls; it is thrilling to see new roots formed and fresh shoots starting to emerge from, essentially, a twig.

Taking cuttings for the first time might seem daunting, but it is immensely exciting. I am one of the lucky ones: it never seemed impossibly difficult for me because my mum was constantly taking cuttings, and they almost always took root. She would plunge prunings from her roses into the ground, pull off sideshoots from penstemons she loved, and turn one argyranthemum into many with what

seemed like miraculous ease. Her garden was tiny, it could never have accommodated the legions of plants she grew, but family, friends, and neighbours were only too happy to benefit from the bounty. Should anyone express admiration for a plant in the garden, a few months later they would be presented with one or two – grown to order! The equipment she used was basic and usually recycled, and most of the decisions she took about when and how to do things were based on intuition and common sense.

It never occurred to my mum that some plants might be difficult to propagate – she would just have a go. When I first tried taking cuttings I had the benefit of her experience, plus a smattering of knowledge I'd garnered from books and magazines. Though I dearly wanted to make new plants just for the joy of it, economic factors – not having enough spare cash to buy the plants that I wanted – were a big incentive too.

In the early days of the garden here at Glebe Cottage, growing food was a priority, but side-by-side with the beans and potatoes little nursery beds were reclaimed from weeds and grass. Here, new plants could enjoy a sojourn, growing big and strong and eventually yielding first-rate cuttings material.

Though many of our first plants were grown from seed, for some subjects cuttings were compulsory. Some plants I wanted could not be grown from seed: coloured-leaf elders, blackcurrants, sages and lavender for a hedge. True, lavender does grow easily from seed, but the results are unpredictable and may result in a higgledy-piggledy edging to a border, growing at different rates, with varied habits and colours. Grow your plants from cuttings all taken from the same parent plant and you are guaranteed uniformity – not always a desired characteristic in gardening, but essential in the case of a lavender hedge. Taking cuttings means propagating vegetatively, so you are in fact creating clones of the parent plant, and as a result you are guaranteed perfect replicas.

Taking stem cuttings also enables us to propagate a vast range of plants that we could not reproduce from seed, perhaps because they are sterile hybrids or don't set seed in our climate. Although it is often thought of as a method used exclusively to increase shrubs, it is also invaluable for making more perennials, and for all those in-between 'sub-shrub' plants that fall between two stools – including lavender, sages and a host of grey- and silver-leaved plants – not to mention climbers of every description. Most shrubs and many perennials have stems that branch and branch again, making excellent material for cuttings. Trees do the same, but cuttings from trees will only be successful in a very few cases. For them, growing from seed or grafting are the two main methods of making new plants.

Cuttings can be taken throughout the growing season, though there is

TENDER CARE
These penstemons are well on their way to becoming bushy new plants, but they will need a sheltered spot until they are properly established.

usually an optimum time for each species. This may be as long as several months, perhaps starting when stems are green and pliable and going on until they are ripe and beginning to become much more rigid. This is the case with penstemons or many cornus species; pelargoniums are also fairly

straightforward, and can be rooted almost year-round. Other plants, though, offer us a much narrower window of opportunity. When taking cuttings from cistus, we collect the material from them at fortnightly intervals, three times from early to midsummer. Although some cuttings from each batch will root, one batch will invariably root far more successfully than the others – which batch succeeds best will vary from year to year. It is that element of unpredictability and the opportunity to experiment that makes growing from cuttings so much fun.

Types of cutting

Making stem cuttings is a propagation method
familiar to many gardeners.

Most of us at one time or another have
tried our hand at removing sideshoots
of perennials or shrubs and nestling
them down in pots or trays. Success
depends on many things: the time of
year, the quality of material, and the
facilities we have at our disposal.
Knowing what is what can up your
chances of success.

To start with, just what are the
differences between all the types
of cuttings – softwood, greenwood,
semi-ripe, ripe, hardwood? No doubt
each category could be split further,
but when it comes down to it, it's
simply a question of how ripe the wood
is. The younger and more supple the
stem is, the faster it will root, but the
more rapidly it will lose moisture and
wilt. Softwood and greenwood are what
they say they are, with greenwood a
little firmer, and these are the cuttings
you take in late spring to early
summer. Semi-ripe cuttings are still

**MORE OR LESS
ABOVE LEFT**
Sometimes you might
pull off whole shoots
with a heel, as on
this *Sambucus nigra*
f. *porphyllophyra*
'Thundercloud'.
RIGHT For other
cuttings, a change in
colour distinguishes
old and new wood,
like on this camellia.

SOFT OR HARD?
Whether a plant roots best from softwood, like this hydrangea (left), or hardwood, like the *Viburnum opulus* (right), always go for strong young shoots that have the best chance of making root.

soft at the tip, but firm at the base, while ripe are firm, but flexible, and typical of later summer.

At one end of the scale, the soft and pliable growth of early summer is raring to go. If we are prepared to cosset our new softwood cuttings by maintaining a moist, but buoyant atmosphere, they will root rapidly and soon become strong young plants that are capable of going it alone. At the other end of the season, hardwood

cuttings taken late in autumn and into winter lose little moisture and remain turgid for a long time. They have no leaves to transpire, and the stems are already so mature they can fend for themselves — but you do have to be patient, because they take months, not weeks, to produce new plants. The method for preparing all stem cuttings from softwood to ripe is much the same — hardwood cuttings and leaf-bud cuttings are a little different.

Choosing and gathering material

Nothing succeeds like success. Those little white roots and fresh green shoots always make you smile.

The huge sense of achievement grows with your plant and reaches its climax as you lower it, grown and lusty, into its planting hole in the garden proper.

Chances are improved enormously if you choose good material from which to take your cuttings. What constitutes 'good material'? The younger and stronger the selected shoots are, the greater the chances of success. It may be a counsel of perfection to suggest preparing the plant well in advance for its future role as a donor, but it is well worth pruning in one year to promote strong, vigorous growth for the next, especially for shrubs. Always use new, young shoots: if you look at bark colour and texture it's easy to distinguish this season's growth. Most of us, though, just grab what is available.

The younger material is, the better it will root, but the underlying age of a shoot is not always obvious. It may be this year's new growth, but its donor parent's age is recorded in every cell,

and cuttings from older plants are far less successful. This effect ripples on through the generations: cuttings from plants grown from cuttings from plants grown from cuttings (and so on) show their age, which is the cumulative age of *all* those preceding generations, or the age that the very first seedling parent would be if it survived. This is the reason that very old cultivars are more difficult to root.

Whether you are raiding perennials, shrubs or subshrubs, there are usually two options. One is to take the cutting from the leading (apical) shoot, which is always the strongest and has the best ability to take root, the other is to use the sideshoots (laterals). You can frequently get both cuttings from the same plant.

CUTTING YOUR MATERIAL

Get everything ready before you start: you don't want to be faffing around searching for pots while your cuttings

SEMI-RIPE CUTTINGS
For most cuttings, just trim below a leaf node and remove the bottom leaves before pushing into a pot up to the next leaf. These physocarpus cuttings were taken from a shoot that was mature, but still bendy – perfect for semi-ripe cuttings.

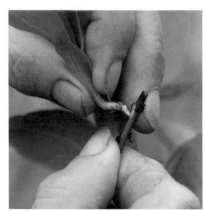

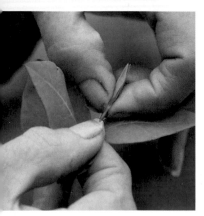

HEELED CUTTINGS
Take a length of stem with sideshoots and gently pull off the shoots with a 'heel' of bark from the main stem. Trim the snag, remove the lower leaves, pinch out the growing tip and insert your cutting into a pot – you can fit several around the edge of one pot.

expire on the touchline. As soon as a cutting is severed from the parent plant, it starts to lose condition: leaves continue to transpire, but none of the moisture is being replaced through the stems from the roots. The softer the wood, the faster cuttings will wilt, so it is vital to be prompt, especially with very sappy shoots. By the same token, the younger and more supple the stem, the faster it will root. Time of day is important. It's worth getting up early to take cuttings when stems are full of sap to give them a flying start and a fighting chance. Try to avoid hot, sunny times, even with Mediterranean plants.

Cut through the stem cleanly, using a sharp knife, scissors or secateurs – fruit pruners are ideal, a sort of hybrid of secateurs and scissors. My penknife is always sharp; Neil showed me how to keep it that way. Whatever you use, it's important not to crush the stem, as this inhibits rooting. My friend Richard Lee, who worked at Rosemoor, could put roots on anything. When he died, his beautiful Tina knife, sharpened every day, was worn thin. He showed me how to safely cut a stem with a sharp knife, aiming to the side of my thumb. Cut directly above a leaf node. This means you leave no stump to rot, and it also encourages sideshoots to grow. If you are detaching a sideshoot with a heel by pulling it away from the main stem, make sure you do not damage the stem: pull down gently, but firmly, and if necessary snip the base of the heel.

Take a manageable number of cuttings at a time: stems will still be there the next day. It is always worth sealing material in a polythene bag as you take it: even then cuttings may wilt by the time you get back to base, but they soon perk up!

CATCH THEM YOUNG
These hydrangeas, grown from cuttings last year, have become leggy. Taking off top growth encourages them to bush out and at the same time provides perfect material for new cuttings.

THE WAITING GAME

When I took my first little batch of cuttings, I was impatient: sages, helichrysum, lavenders, were all levered gently, but prematurely, from their resting place for a half-time inspection. I soon learned to leave them alone. Cuttings need time to produce self-supporting roots, and however tempting it may be to check on progress, they should be given optimum conditions, then left to their own devices.

Soon enough the cuttings themselves will let you know they have rooted. When you first take them, they will flag suddenly – they have undergone severe trauma being severed from their roots. At this stage all you can do is to replace some of the precious moisture that they can no longer take up from their roots. Keep the compost moist and spray with a fine mist (the little plastic sprays used to damp down ironing or house plants are ideal).

When your cuttings begin to show signs of growth, making new leaves and looking healthy and sprightly, gently turn out your pot or one of the modules to inspect the roots. If they are well formed, put them back and prepare everything to pot them up. Just like people, cuttings vary in the time they take to get going. One great advantage of giving cuttings their own space in a module tray is that each little plant can have individual treatment and can go up a rung when it is ready. In contrast, when several cuttings are growing together in one pot, some may be really well-rooted, while their neighbours may only just have started to make roots, and you are forced to take a middle-of-the-road practice, potting them all up at the same time, ready or not.

If they have been under glass, a weaning-off period after potting up will make for sturdy plants. Put them back where they came from for a week and then gradually introduce them to life outside (see p.214).

Penstemon

Softwood cuttings aren't just for shrubs:

Taking cuttings is not only a great way to increase your stock, it is also an insurance for plants that are not reliably hardy. Among the most rewarding herbaceous plants to grow from cuttings are large-flowered penstemon cultivars. Everyone has their own rules: I use clay pots because their porous walls breathe and cuttings root readily in contact with clay (after all, it's fired earth). And I always nip out the tip of the stem to encourage the cutting to put down roots rather than attempt to keep growing upwards. Others leave it, believing this will promote rooting. A plant biologist told me recently there were arguments for and against both methods. I will carry on with my practice because it works well – my cuttings root and my plants bush out.

1

2

3

4

5

6

try them for herbaceous perennials that may not survive winter.

7

8

9

10

11

PLANT TO PLANT

1 Collect material from the plant early in the day.

2 Cut just above a pair of leaves to encourage lateral shoots to form.

3 Trim the apical shoot back to a pair of leaves.

4 Remove the lower leaves to give a bare 4–5cm (1¾–2in) of stem, and insert around the edge of a clay pot – about six to a 15cm (6in) pot.

5 Finish with a layer of grit to retain moisture and keep down weeds, and water thoroughly.

6 After a month or so in a warm, bright place, cuttings have made good roots.

7 Pull apart the rooted cuttings.

8 Pot on each one individually.

9 Put in a sheltered place to grow on.

10 After another month, they make fine little plants.

11 Established plants are put in their final places.

Hardwood cuttings

We take so many cuttings here at Glebe Cottage that sometimes plants are queuing up to be propagated.

Although there are optimum times for taking cuttings, for most plants there is generally a leeway of a few weeks during summer and early autumn where softwood and semi-ripe cuttings will root. But as the leaves begin to flutter from trees and shrubs, we start to concentrate our attention on one of the easiest and most relaxed methods of propagation: hardwood cuttings.

One of the problems with most cuttings is making sure that they don't lose so much moisture through transpiration that they shrivel before they have had time to root. But with hardwood cuttings this is never a concern, because they are best taken at leaf fall. They are among the most successful cuttings. What's more, although they take a long time to root, they get on with it on their own.

What can you grow from hardwood cuttings? Almost any shrub, including fruit bushes. Plants that make good, strong, straight stems work best; blackcurrants, shrub roses, viburnum, physocarpus and cornus are good examples, but there is little to lose by experimenting with almost anything. Spirea, hydrangea, forsythia, weigela and deutzia are certainly worth trying.

The major point to remember is that stems should be mature. The name hardwood cuttings speaks for itself: stems should not be bendy or soft. They should have finished growing for the season, and their leaves should be falling or about to fall. Even if a stem still has a few leaves adhering to it, if

they fall off when you run your hand down the stem it should make ideal material. If you miss this moment, don't panic – hardwood cuttings can be taken at any time during the dormant season, it's just that the optimum time is at leaf fall. Ideally, the shrub should have been pruned well during the previous winter or early spring to help it make vigorous new growth, but this is not essential. Try taking a whole stem from close to the ground: you should get several cuttings from it, although the cuttings closest to the base will root most easily.

PREPARING A TRENCH

For many gardeners, setting aside part of a coldframe for hardwood cuttings is not an option. Even if you are lucky enough to have such a facility, there are many other more versatile functions it can fulfil, since hardwood

LINED UP
Material for
hardwood cuttings
is best collected at
leaf fall. Line a slit
trench with grit and
insert the cuttings
vertically, with just
a couple of inches
protruding when the
soil is returned.

the soil to a fork's depth, firm gently, then make a trench (or a series of little trenches if that fits in with the space you have) by pushing a spade into the soil vertically and pulling it forward slightly. Line the bottom of the trench with coarse grit or sand for drainage.

TAKING THE CUTTINGS

At one time hardwood cuttings would be taken up to 45cm (18in) long, but experimenting has shown that shorter cuttings about 15–20cm (6–8in) work best. You should be able to make several cuttings of this length from one long stem, each with a sloping cut above a bud (or node) at the top and a cut straight across the stem, again just below a bud, at the bottom. Line up the cuttings along the back of the trench with two or three buds protruding above the level of the soil and fill in, firming the soil gently. Water well if the soil is dry, label, then leave alone. Apart from checking the watering occasionally and refirming the soil if frost has lifted the cuttings, there is no other work involved. In spring new shoots will appear and in the autumn, when your new plants lose their leaves, they can be lifted and either transplanted into their new homes or potted as swaps or gifts. By this time they will have formed a strong root system and made viable plants.

The other time you can strike hardwood cuttings is just before leaf burst in the spring, but the material is still collected at leaf fall in the same way and stored in damp sand till the spring. Like several other practices this had as much to do with historical convenience as efficacy: in big estate gardens, late winter would have been the quietest time to fit it in.

cuttings occupy their space for a whole year. My mum had a small bed close to the kitchen where she grew a few vegetables and rooted rose prunings and stems of shrubs both from her own garden and from friends and family (quite a few of hers came from Glebe Cottage, and vice versa). Her method was not particularly refined, she would just stick them in the earth and hope for the best. Most of them rooted though.

Choose a sheltered corner with well-drained soil where you know the cuttings can remain undisturbed until it's time to transplant them. Fork over

Leaf-bud cuttings

A few plants, including such popular genera as clematis and camellias, lend themselves to this economical technique.

The main difference between leaf-bud and normal stem cuttings is that whereas most cuttings are taken directly below a node, leaf-bud cuttings are taken internodally, so the base of the cutting is between two nodes. They can be taken at the same time as other stem cuttings, with any kind of wood, and the same rules apply when it comes to choosing material: strong, young and free from disease. The same rules apply to collecting too: cut the stem cleanly, directly above a leaf node to guard against rot and encourage branching, and gather material as quickly as possible, putting it into a plastic bag.

To prepare the cuttings, make a clean, sloping cut just above a leaf node at the top, and another horizontal cut about 4cm (1¾in) below the leaf node. Each cutting needs a leaf-bud, which will be in the axil of the top leaf. This will become the new stem of the plant the cutting will make. The leaf will provide the cutting with enough sustenance to keep it going until its new roots and leaves can support it. The piece of stem under the leaf node is where the new roots will form, so when the cuttings are nestled into their compost the whole of the stem must be under the surface, with the leaf and its leaf-bud sitting on the surface. Use a chopstick or dibber if the stem is soft, but if it is harder just push

CAMELLIA CUTTINGS
Cut a main stem back to just above a leaf or pair of leaves. Young wood makes the best cuttings, it is a far more vibrant colour than the old wood. Trim the top of your cuting just above a leaf bud, making a cut that slopes away from it. Cut straight across the stem about 4cm (1¾in) below the leaf for the bottom of the cutting. Insert all the cuttings right up to the leaf. They will root from the buried stem and shoot from the embryonic bud.

it gently but firmly into the compost. This will abrade the stem slightly, which helps stimulate it to produce new roots. With subjects that are difficult to root it is sometimes worth wounding the stem by taking a fine sliver off the lowest two-thirds of it. Cuttings can be placed around the edge of a pot or in separate modules or even individual pots if they have big leaves. Mahonias take well from leaf-bud cuttings, and I give each cutting of *M. japonica*, with its big, shiny evergreen leaves, its own pot.

Some mahonias have prickly leaves, as I discovered when I went to learn a bit about propagation with Richard Lee more than 20 years ago. It was late October and, according to Richard, at Rosemoor, this was the ideal time to propagate mahonias, their even more prickly cousins berberis and hollies, of which Rosemoor has a comprehensive collection. By the end of a fortnight my hands were scratched and scraped and I was praying we would move on to more soft-leaved, tender subjects. It was a great education though: the only way to learn is by doing it, and how lucky I was to have such expert guidance.

MAHONIA CUTTINGS
Big-leaved subjects, like mahonia, can seem like awkward customers, but they root well from these bud cuttings. If the leaf is too big to be accommodated, it can be cut down or rolled up and secured with a rubber band.

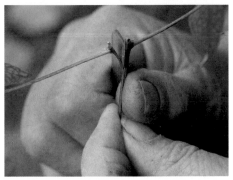

CLEMATIS

These climbers are reasonably easy to propagate from leaf-bud cuttings – you can even double your money by splitting each cutting vertically in two, down the stem. Help these cuttings, which can be quite brittle, into the compost with a chopstick or dibber.

Some plants propagated in this way – clematis is probably the most notable example – make twin pairs of opposite buds along their stems, and you can take two cuttings for the price of one. Take material in just the same way as for cuttings with a single bud, but when you trim above the buds, make a horizontal cut. Either lay the cutting down with leaves at either side and cut the stem between them with a very sharp knife or hold the cutting upright and cut through it from above. Each cutting can be planted individually. If you are too nervous to try this dissection, the whole cutting can be plunged into the compost up to its leaf buds in the usual way, removing one of the leaves if necessary.

The after-care procedure is just the same as for other stem cuttings, dressing the surface with grit, watering thoroughly to settle cuttings in and spraying frequently to reduce transpiration. Eventually the leaf-bud will start to grow and the infant shoot will emerge from it. When roots are properly developed, cuttings can be potted on in the same way.

Tender cuttings

During winter, gardening work slows to a standstill, especially when typical wintry conditions prevail.

Outdoor jobs – usually the energetic digging over of a border or building of a new wall – are about all that can be tackled, and only in clement weather. But providing you have a greenhouse, or at least a bright porch or windowsill, you can carry on propagating even through the wettest and coldest days. Light is in short supply, and sowing seed can be a thankless task, but taking cuttings from tender plants is an ideal occupation, and at this most barren time of year it can yield a wealth of new plants.

Even more important than the right conditions to help cuttings root is the material to be used. Most of us bring things in for the winter, and amongst these hoards of tender plants there are almost bound to be subjects that are just begging to be increased. Suitable material can be purloined from salvias,

aeoniums and pelargoniums, from plectranthus, verbenas and gazanias. As light increases in the new year, the growth of the cuttings reflects the surge of spring. Tender plants that were cut back or trimmed for their sojourn under glass often provide ideal material. They will continue to grow right through the winter, especially if their space is heated or even just frost-free. Depriving them of excess sideshoots or shortening top growth prevents them becoming drawn and attenuated in lower light levels, so both parties benefit.

Tender perennials are in the garden ascendant nowadays. The fashion for the exotic has resulted in more and more of us using them, so we need to know how to overwinter them and increase them. Most gardeners have taken cuttings from pelargonium as

SALVIA CUTTINGS
Any stems that are solid, rather than hollow, should root easily. Cut just below a leaf node, remove the lower leaves and nip out the top. Insert cuttings around the edge of a clay pot.

an insurance policy: some of the subjects more recently in vogue are just as easy or even more straightforward.

SALVIAS AND RELATED PLANTS

Salvias, from Central America, have become *de rigeur* for the late-summer garden. At Glebe Cottage we have a rolling programme for propagating these exciting plants. A batch grown from cuttings taken in the summer will yield a mass of suitable material for further cuttings during the winter. Either leading shoots or sideshoots can be used. With a sharp knife sever pieces 10–15cm (4–6in) long, just above a bud so that the parent plant can heal the wound without rot setting in. Trim the cuttings below the lowest leaf node, nip out their growing tips and remove the basal leaves and any other large leaves. There is a fine balance between retaining enough leaf-surface to keep the cutting alive and leaving too much, resulting in excessive loss of water through transpiration. All these soft cuttings wilt immediately after they have been taken, but recover rapidly.

Excessive watering in a bid to resuscitate them is counter-productive and can result in rotting. Ideally use a loam-based compost with extra grit. Although any proprietary potting compost will do, loam-based is best for winter cuttings because it contains more nutrients, which will help sustain the rooted cutting before it is potted on.

Fill the pot loosely up to its brim. Dibble the cuttings in around the edge, up to the first untrimmed leaves. Winter cuttings can be quite soft, so ease them in gently, making a hole first with a chopstick or dibber. Half a dozen cuttings can be comfortably accommodated around the edge of a 15cm (6in) pot, as for all other cuttings, terracotta is best.

When cuttings have rooted strongly it is probably best to turf them out of the pot and carefully separate the rootballs. Salvias and other members of the labiate and lamium families – mints, agastache, plectranthus and coleus – develop good strong root systems and can take this sort of manhandling.

FAST PROGRESS
When cuttings are well rooted, tip out the whole pot, separate carefully and pot up individually. These salvia cuttings rooted within two months and had made so much growth by the time we came to pot them up individually that we took more cuttings from them before we moved them on.

SEPARATE POTS FOR SUCCULENTS

There are other plants that benefit from having their own private space in which to take root. Succulents, for example, have strong structures above the ground, but their roots are often fragile. If they are grown in separate modules or pots, the transition to bigger pots once they have successfully rooted is not so traumatic as being torn away and separated (however carefully) from others of their ilk.

Aeonium 'Zwartkop' has become a 'must-have' plant in recent years. Its perfectly symmetrical rosettes of fleshy, deep bronze leaves are easy to accommodate when the plant is young, but as it ages it can become ungainly, developing into a tree-like structure with clusters of rosettes atop tall stems. A native of Madeira and the Canary Islands, it cannot stand extreme cold and must be brought under cover for the winter and kept frost-free. Whilst it is indoors it can be propagated easily and very effectively. Steel yourself, and with a sharp knife cut off as many rosettes as you need

with a piece of stem 8–10cm (3–4in) long. Remove any shrivelled leaves from the underside of the rosette (all succulents constantly shed old leaves as their rosettes expand). Loosely fill a module tray or individual small pots with crunchy compost and press the stem firmly into it until the rosette is all but sitting on the rim. Sprinkle extra grit over the surface. Give one thorough watering and don't water again until the compost starts to dry out. Succulents dislike sitting in damp conditions, and if the cutting has to resort to drawing on its own water reserves in the leaves it will be prompted into making roots.

An aeonium plundered for cuttings can present a forlorn picture, but new rosettes will appear from dormant leaf buds wherever the stem is cut back to. One of our tallest and most handsome aeoniums blew over in a storm and snapped off just above ground level. We trimmed it back cleanly and made a tray full of cuttings. The old denuded stem was carelessly shoved into a pot of old compost and has spontaneously

AEONIUM CUTTINGS
Winter is a good time to propagate succulents. Take a chunky piece of stem that terminates in a large rosette, remove any old leaves and plunge it to the hilt in modules.

sprouted a whole new generation of rosettes. Eventually these too will be removed and potted into trays. Care is simple: pot on carefully, and keep in as bright and sunny a position as possible. When the weather warms up and the threat of frost is over, plants can be put out for the season.

SOUTH AFRICAN DAISIES

Many tender plants in the repertoire are from South Africa. Gazanias and osteospermum, both daisies, are widely used in our gardens and valued for their long and prolific flowering. Some osteospermum will stand a few degrees of frost – even when their evergreen, branching top-growth is badly damaged by cold it may protect the base of the stems, and if they are cut back hard in late spring they will often spring into life from the base. In gazanias however, all the growth is basal, and if the plant is frosted it will die. There are some especially good named gazania cultivars, and even in a batch grown from seed there may be one or two plants worth increasing vegetatively.

The first time I propagated a gazania I was completely stumped, but on closer examination it was clear that the only method was to detach whole sideshoots right at the base and gently ease them away from the main stem just above the root. This can be done from the plant growing in the garden or when it is lifted and brought inside before potting up. Carefully strip a few of the longest leaves from the base of the shoot and, even though it seems there will be little left, get finger and thumb into the heart of the shoot and nip out the centre. If the leaves are very long they can be cut in half, but this is really best avoided with winter cuttings since severed leaves sometimes rot. As with the other cuttings, these shoots should be pushed into sharply drained compost and potted on when rooted.

So, when you fling open the greenhouse doors as spring turns to summer, there will be a host of new plants ready to embellish containers and exotic planting schemes, all made during the dire days of winter.

GROWING UP
In module or cell trays you can space cuttings out so their leaves don't crowd each other. When they are growing strongly, move each one up to its own pot of gritty compost.

Basal cuttings

Nurserymen's catalogues of yesteryear listed scores, sometimes hundreds of named varieties of perennials like phlox, delphiniums and lupins.

Amongst the multitude of mouth-watering plants a few shone forth especially brightly from the pages. *Ostrowskia magnifica*, a campanula described by Graham Stuart Thomas as 'unbelievable', was listed in the 2009 *RHS Plant Finder* as being stocked by only three nurseries, yet up to the Second World War it was widely available. Back then, labour was cheap and nursery staff were skilled in the production of new plants from basal cuttings. Nowadays, such production would be uneconomic, so plants like this have all but vanished in cultivation: on the sun-drenched slopes of its native Turkestan it undoubtedly seeds itself, but our summers are seldom hot enough for it to set seed, so basal cuttings are the only option.

Of course, this ostrowskia is unusual: the great majority of species perennials can be grown from seed. But increasing cultivars or selections is more problematic. These are the plants that friends and neighbours *ooh* and *aah* over. They are valued for a unique characteristic – the depth of colour of their flowers, perhaps, or the particular patterning on their leaves – and too often they lose the very qualities for which they are valued if grown from seed, reverting to the straight species.

Quite a number of them are incapable of setting seed, either because they have doubled, sterile flowers or because they are genetically sterile hybrids.

There is the alternative of division (see pp.162–189) but not every plant is amenable to being pulled apart; some, like delphiniums, positively resent it, and others simply cannot be divided, because they they have tap roots, a solid, impenetrable crown or new shoots that arise from a woody rootstock. How do you explain to friends that it is impossible to slice off a chunk, when they look so longingly? Taking basal cuttings may be the way to keep them happy and increase your own display. New plants can be made from basal cuttings without damaging the plant, and with amazing alacrity compared with growing from seed, providing it is the right time of year.

SUITABLE CANDIDATES
You can try this technique with any perennial that makes multiple stems from one crown. Classic examples that lend themselves to it are lupins and delphiniums; this is the time-honoured method used by estate gardeners to make more of their most prestigious perennials and fill the huge herbaceous borders for which large properties

were, and still are, renowned. Not only did it help increase numbers, but it ensured that there was always fresh, healthy stock to take over when old plants had to be hoiked out. If you have a favourite delphinium, whether it is a named variety or one you grew from seed, this is how to make more of it.

It is possible to prise lumps of border phlox apart, but it is far more fruitful to increase them from basal cuttings. Proud owners of special phlox who want to make more need to give their plants an extra treat during the previous growing season, to help them produce strong, new growth in the spring. Any balanced fertilizer will do; we use an organic liquid feed. Avoid high-nitrogen feeds, as they produce sappy growth. As with other cuttings, take them as early in the day as possible, when growth is at its most turgid.

Phlox have fairly solid stems, but those of lupins, delphiniums and most labiates (plants related to mints, with their distinctive, two-lipped flowers) become hollow very quickly. Only solid stems will root, so once this happens they are useless as source material for basal cuttings. *Lamium orvala* is a case in point. This is such a sumptuous and exciting plant that you could never have too much, and at Glebe Cottage

PHLOX CUTTINGS
Basal cuttings should be taken from the first flush of new growth. Sever shoots as low down as possible, sometimes even below soil level.

we have a particularly striking clone with large, richly coloured flowers. Each year we intend to take basal cuttings, but spring is a busy time and almost invariably it is overlooked until it is too late. Its square stems must be taken where they emanate from the root and when absolutely solid.

TAKING BASAL CUTTINGS

When shoots are 10–12cm (4–5in) high, sever them at the base as close to the rootstock as possible with your ever-sharp knife. Strip the leaves from the bottom half of the stem with a sharp knife or a thumbnail. Pinch out the growing tip to concentrate efforts into forming roots, rather than heading for the sky, and pre-empt any attempts the cutting might make to flower – except labiates that flower in whorls all down the stems, so you may need to remove flower buds as they develop.

You can use pots, deep seed trays or module trays, but again, terracotta pots are the most sympathetic. The compost needs to be open, yet moisture-retentive, and capable of sustaining growth for a while, since the cuttings take some time to root. Cuttings might root easily in

grit or vermiculite, but these offer no nutrients to new growth. A proprietary cuttings compost will do, or you can make your own from equal parts sieved coir or composted bark, sharp grit, and sterilized loam. Plunge the cuttings around the edge of the pot, up to the base of the lowest leaves. Give them enough space to grow and develop. As ever, finish off with a layer of moisture-retaining, weed-inhibiting grit, water thoroughly and put into a warm, light spot, out of direct sunlight. These cuttings root most speedily with bottom heat, but it is not essential to success. Because they are liable to wilt at first, it is important to spray the leaves with a fine mist from time to time to replace moisture lost through transpiration.

In all cases, once the cuttings have become plants they should be potted on without delay. New growth usually indicates that rooting has occurred. Turn out gently and separate, keeping as much compost as possible on and around the roots. Pot individually into a more substantial compost, water well and as soon as plants are properly established pot on again or plant out in a well-prepared site.

MOVE FAST
Cuttings from such young material as these phlox (above) or dahlias (right) may wilt quickly, but providing they are potted rapidly and watered well they will soon recover. Maintain good humidity with all cuttings, but especially these.

Dahlia These tender plants are a classic candidate for basal

As overwintered plants are re-potted and watered in the spring, or as those left in the garden begin to come back to life, they produce a plethora of new shoots that are perfect for poaching. An older plant should send out several new shoots, and a number of cuttings could be taken without doing it any harm. In fact, providing it is left with seven or so robust, evenly spaced shoots, taking cuttings will allow the plant to concentrate all its growing potential into the shoots you have left, making their stems, leaves and flowers all the more luxuriant.

1

2

3

4

5

cuttings, taken as the plants begin to grow in the spring.

6

7

8

9

10

TAKING CUTTINGS

1 When shoots are a handspan tall and still solid, expose the crown.

2 Slide a sharp knife between a strong shoot and the crown of the plant, cutting without damaging the tubers.

3 An older plant will yield a handful of strong cuttings.

4 Push the cuttings firmly around the edge of the pot. Use a free-draining, substantial mix.

5 Grit the surface for drainage and protection.

6 Drench the pot to settle the compost.

7 When roots appear from the hole in the base of the pot, re-pot each cutting.

8 Knock cuttings out of the pot and separate carefully.

9 Lower plants into their own pots, taking care of fragile roots.

10 Finish with grit and grow on.

NATURAL GENIUS

The ability to root from a fragment of stem keeps some plants going.

DIANTHUS

All manner of plants can be increased from cuttings, and there are groups of plants that have been traditionally increased and passed on by our gardening forebears in this way over centuries. Amongst the most popular of these are garden pinks, mentioned in one of the first gardening books (using the term loosely!), Gerard's *Herbal*, first published in 1597.

It is the clove scent of old varieties and their fringed or formal flower shapes that appeal. For the amateur gardener, propagation from cuttings is the only way to keep these antique plants going. One we have from an old garden, which we call 'Elizabethan', has rounded, white petals edged with dark maroon and a splodge of the same colour in the centre. Its perfume on a hot afternoon in early summer is almost overwhelming.

Cuttings ensure the survival of new varieties too. We once had the lovely deep pink *Dianthus* 'Waithman Beauty'. Although it was dead-headed regularly, one seedhead ripened and the seed was sown. One of the seedlings was an exceptional plant. 'Glebe Cottage White' has semi-double flowers composed of frilly white petals, with centres that turn slightly pink in warm weather. It spreads well, but has a very neat habit and, unusually, it flowers perpetually from late spring until the autumn. No matter how many seeds we sow, there will never be another 'Glebe Cottage White', but we can make as many as we like thanks to the

OBLIGING PLANTS
Whether they are cut, pulled off with a heel or simply pulled out with finger and thumb, 'pipings' from dianthus are easy and responsive.

eagerness of wounded stems to put out new roots. What is more, people who take home a plant or just a few pieces can do just the same, and more people can share in a special plant.

Dianthus don't last very long in our heavy acid clay, but we love them, so we take lots of cuttings and renew our plants regularly, adding lime rubble and grit at planting time. Midsummer is the ideal time for this operation, but even cuttings taken in the depths of winter will root successfully, as long as it is not freezing and they are given warm, sheltered quarters. Use short tips of the leading shoots, strong basal shoots or sideshoots pulled away from the plant with a heel.

In all cases remove the bottom leaves and nip out the growing tip to encourage good rooting and bushy plants. Because dianthus have jointed stems, you can pull out a bit of the leading shoot at one of the joints by holding the stem under a joint and gently pulling with the other hand above the joint. These cuttings, the traditional way to increase pinks and carnations, are called pipings. Dibble them around the edge of a clay pot, water them well, and stand them somewhere warm. When the cuttings begin to put roots through the bottom of the pot, they are ready to separate. Pot them individually and put them out in the garden as soon as possible.

The big picture

Growing plants from cuttings is something every gardener can accomplish. In a matter of months we can all produce batches of plants to make our designs achievable, with little or no expense. Every year our dahlias are carried out to swell the ranks of the autumn display and we are reminded of how they were once started from cuttings and now provide material themselves for new plants.

LEAVES

 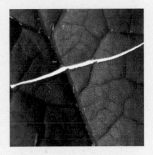  

LIFE FROM LEAVES
Although usually used for houseplants, this method can also be employed to make more of garden perennials and bulbs, like this splendid *Eucomis* 'Sparkling Burgundy'.

I once bought a plant of the double lady's smock, *Cardamine pratensis* 'Flore Pleno'. Having searched for it for ages, I was thrilled to plant it in a suitably damp and fertile corner where it flourished. So just imagine my horror months later when I discovered that the whole plant had been decimated by pigeons. The damage seemed to have been done for some time, and severed leaves from my treasure were lying around on the ground. They were all still green, though, and on closer examination it was clear that each leaf had made roots and was well on the way to becoming a new plant.

The rooted leaves were duly lifted and potted up, and the next year, *Cardamine pratensis* 'Flore Pleno' was a prominent feature of Glebe Cottage Plants' own catalogue. My mum used to increase her favourite streptocarpus and African violets (*Saintpaulia*) from leaf cuttings, which I found a fascinating process, but it had never occurred to me that some hardy perennials could be started this way.

The propagation of any plant from a leaf is miraculous, and it doesn't have to wait until spring. Winter can be a frustrating time for those who love to propagate their own plants. Cuttings from tender plants are usually taken when they are brought under cover at the very start of their winter sojourn. There is then a lag until seed sowing can begin in earnest, when the days are longer and the earth begins to stir. But some houseplants offer salvation to frustrated gardeners with the winter propagation blues. Streptocarpus, African violets and begonias are among the most popular and widely grown of houseplants and all of them are easily increased by leaf cuttings. The best bit is that since they are all evergreen perennials from tropical climes that keep growing all year, cuttings can be taken from them successfully at any time. Although leaf production is reduced in the winter, due to shorter day length and lower temperatures, all the plants in question will continue to yield fully grown leaves throughout the colder months, albeit at a slower rate.

For those who have never taken leaf cuttings, it may seem almost irrational that new plants could regenerate simply from leaves, or even parts of leaves, but some plants have developed the ability to grow new plantlets from the broken or cut edges of their leaves, and it is this ability that we exploit. To help stimulate it, gardeners have developed several different techniques; all of them straightforward. In general, the idea with all leaf cuttings is to sever the leaf and ensure that at least one cut edge is in contact with the compost. Plantlets will be produced along this cut edge, where the lateral veins or the central midrib are in contact with the soil.

With each different method the raw material (the leaf) must be mature, but as youthful as possible. Maturity is important; a leaf that is not fully grown will continue to grow to maturity before it starts to make new plantlets, and the longer the cuttings take to root, the greater the likelihood that they will simply rot away rather than growing. Another safeguard against loss, especially important during the winter months and when propagating organically, is to ensure that all the implements used are perfectly clean and sterile.

What the pigeons showed me how to do all those years ago I can now do for myself. Anyone can have cardamines galore with the help of nothing more than a tray of compost and a sharp knife – in fact you don't even need the sharp knife for cardamines, just your fingernails will do. To see a batch of young, healthy plants ready to go out into the garden created from nothing more than a few leaves is truly rewarding, and to give away streptocarpus or African violets to friends or family who admired the original plant gives you a special satisfaction.

All leaf cuttings need a humid atmosphere and will root most successfully if they have basal heat. They need adequate light to make enough food for the developing baby plantlets, but they resent direct sunlight, which scorches and shrivels them. The three houseplants that are most widely propagated from their leaves – saintpaulia, streptocarpus and begonia – are all creatures of the shade in their native habitats, and they should never be kept in full sun.

The only bits of equipment you need for this method of propagation are a wickedly sharp blade and a sheet of glass. Why a sheet of glass? With a wickedly sharp blade, the hard surface will give you a far cleaner cut than a wooden or plastic chopping board, and a crisp, unbruised edge makes for a more successful cutting. Other hard surfaces might also do, but since the sheet of glass can be popped over the cuttings to keep up humidity once they are in the compost, why not put it to use twice over? Glass will take the edge off your blade more quickly than other surfaces, though, so be prepared to sharpen it frequently as you work.

THE KINDEST CUT
A fully developed leaf can be taken from a *Begonia rex* and cut into squares that look like a ready-to-assemble mosaic kit.

Vein and midrib cuttings

There are two simple ways of increasing streptocarpus. These are also good for other plants that have either a pronounced midrib and veins or a long, strap-like leaf.

VEIN CUTTINGS
It couldn't be any easier. Sever a fully developed leaf from a streptocarpus and slice either side of the midrib. Then simply ease the two halves gently into a tray of well watered potting compost.

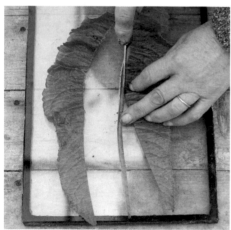

Before you remove your leaf from the mother plant, fill a seed tray with cuttings compost. This should contain plenty of drainage material, such as coarse grit or vermiculite. Adding these materials to regular potting compost is the easiest way to get the right consistency and, as long as you make sure that all the ingredients are

thoroughly mixed, you can create an excellent rooting medium.

Seed trays are the ideal container for rooting most leaf cuttings. Whatever you choose, it needs to have a big surface area to accommodate long leaves, such as streptocarpus, but it can be fairly shallow. Fill and water well with a fine rose to ensure the

MIDRIB CUTTINGS
The leaf can be cut, equally effectively, into transverse sections. Push each slice into compost, leaving enough room between them for plantlets to develop.

compost is settled, then allow it to drain while you cut up your leaf.

Detach a fully mature leaf from the plant at its base. Use a sharp knife or blade and sever the leaf as low down as possible, even if this means removing some compost from the top of the pot. Avoid taking any leaves that have been fully expanded for a long time because

they will take longer to root. Lay the leaf on a sheet of glass and cut it, using a very sharp craft knife, penknife or even a scalpel for the best results.

LATERAL VEIN CUTTINGS
The very simplest way of propagating leaves like streptocarpus is to make lateral vein cuttings. To do this, slice

each leaf into two sections down the middle. Lay the leaf onto a sheet of glass and make two long cuts, one on either side of the central midrib. Make two long, shallow trenches in the compost with a ruler or plant label, and push the cut edge into the trench so that the cutting stands upright and doesn't flop. With this method, plantlets should appear all along the cut edge of each section. When they are large enough for an independent existence, and have made a proper root system, they should be detached carefully from the parent leaf and potted individually. Or the whole leaf can be cut into sections, each with its little plantlet still attached, and then potted up.

MIDRIB OR TRANSVERSE CUTTINGS

You can also make transverse cuttings. Simply cut the leaf straight across, through the midrib, into sections 2.5–5cm (1–2in) wide. Each mature leaf from a streptocarpus should yield at least five pieces, and possibly as many as ten.

EUCOMIS SECTIONS
We have several splendid specimens of *Eucomis* 'Sparkling Burgundy', but we can always use more. Cut a mature leaf into sections, push into compost and settle it with a layer of grit.

Make a shallow trench with a plant label and insert the basal cut edge into it. Repeat the process, allowing enough space between the cuttings (a couple of fingers' width) for new roots to develop. Push the cuttings down just enough to ensure they stand upright and do not wobble over. Label and put in a warm, humid place – a propagator with bottom heat or somewhere similar.

Infant plants will be large enough to lift and pot up separately between one and two months after the cuttings are taken. Pot them up individually, with the piece of leaf from which they have grown still attached. This will go on feeding the plantlet before eventually shrivelling up.

MONOCOT LEAVES

Snowdrops (*Galanthus*), hyacinths, leucojum and scillas can sometimes be persuaded to produce new plantlets from slices of their leaves cut at right angles to the leaf veins, the same method as for midrib cuttings. Once again, mature leaves and scrupulous hygiene are all important, and humid conditions and bottom heat promote plant formation. When you have something really special, like the splendid eucomis cultivar 'Sparkling Burgundy', it is well worth sacrificing one of its glorious deep crimson leaves in an attempt to make new plants. Its leaves are long too, so there are plenty of chances; you may easily make ten cuttings from one leaf. This particular plant carries 'Plant Breeders' Rights' certification, which means that you may not sell plants you have grown, but there is nothing to stop you making as many as you like – or as you can – for your own garden. Good luck.

Leaf slashing and squares

Begonias have been high on the top-ten house plants list for some time and, thanks to several excellent nurseries, new varieties are steadily being introduced.

LEAF SLASHING
Just slashing veins and laying a leaf on damp compost is enough to instigate rooting. The veins are most clearly visible on the back of the leaf. Short staples help to keep the cuts in contact with the compost, and a glass lid maintains humidity.

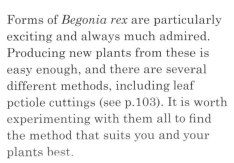

Forms of *Begonia rex* are particularly exciting and always much admired. Producing new plants from these is easy enough, and there are several different methods, including leaf petiole cuttings (see p.103). It is worth experimenting with them all to find the method that suits you and your plants best.

LEAF SLASHING

Leaf slashing is the easiest method. Though it sounds brutal, after the initial surgery it is just a question of letting nature take its course. Fill a seed tray with cutting compost, water thoroughly and allow to drain. Use a sharp knife or blade to cut a recently matured leaf cleanly from the plant.

LEAF SQUARES
Cut off the stem of a fully-grown leaf, slice it evenly into long strips, then cut them into squares.

Place the leaf upside down on a sheet of glass. Make a slit about 2cm (¾in) long across one of the main veins and then continue to make similar slits across other veins over the entire surface of the leaf, leaving about 2.5cm (1in) between each cut. The reason for cutting on the back of the leaf is that the veins are more pronounced on this side, and it is far easier to see what you are doing.

Turn the leaf over again, so it is topside up, and press it gently onto the surface of the tray of compost. If it will not lie completely flush, help it to make intimate contact by pinning it down with short, wire staples – florist's wire can be easily cut into convenient pieces

and is very malleable – or use little stones to weight the leaf down, but make sure they are washed first so that they do not introduce disease.

Label and cover with a sheet of glass. This creates a close, humid atmosphere while allowing in the light needed for new plantlets to form. There should be at least 1cm (½in) between the top of the leaf and the glass. A heated bench or propagating case is the ideal environment for plantlet formation, although a cover is unnecessary since the glass takes its place. Gradually, plantlets will appear from each slash. At this stage they are best left in situ until they can be easily handled. When small plants have developed, remove the glass and give the plants an occasional spray with water to retain humidity. A weak organic liquid feed can also be applied to promote growth.

LEAF SQUARES

Some leaves just refuse to lie down flat: many begonias have leaves with a rough or puckered surface, and it is impossible to persuade them to make close contact with the compost. Cutting leaf squares may improve the chances with recalcitrant leaves, and just takes leaf slashing one stage further.

Take a fully expanded leaf from the plant. Remove the stem and put the leaf face down onto a sheet of glass. Using a straight edge, or a narrow piece of wood, cut the whole leaf from top to bottom into strips. Turn the glass around 90 degrees and cut across again so that you end up with the whole leaf cut into 2cm (¾in) squares.

Push the squares into a tray of watered and drained compost just enough to hold them upright, or, if the

VERTICAL SQUARES OPPOSITE The leaves of some begonias, like the aptly named 'Escargot', may not lie flat. Squares are instead pushed upright into damp compost. A wire frame provides support for their own mini-greenhouse.

surfaces of the squares are not too undulating, lay them flat on top of the compost. When the tray is full make a framework over it with florists wire and put the whole caboodle into a clear polythene bag. An inverted punnet or other clear plastic covering is less fiddly, especially if you match tray and cover before you start. Again, bottom

heat speeds up rooting, but it is important to retain humidity without getting things too wet. An occasional spray with a fine mist is a cheap alternative to a professional mist unit. Regularly checking on the humidity gives you the opportunity to keep an eye on your cuttings. Each one should eventually develop a small plant.

LEAF PETIOLE CUTTINGS

LEAF PETIOLES A whole saintpaulia leaf is severed and its petiole shortened. The petiole is pushed into compost up to the leaf base.

Saintpaulias do not have a pronounced midrib, nonetheless they are very easy to root from leaves. Once again, it is important to select leaves that have recently reached maturity on a healthy, vigorous plant. Sever the leaf stalk at ground level and shorten it to about 5cm (2in) long.

With a dibber, make a hole in the compost and lower the leaf stem into it, right up to the base of the leaf. It is best to use pots or deep modules for leaf petiole cuttings, since their stems need some depth of compost, and several cuttings can sit round the edge of a pot. Firm gently and

repeat the process with as many leaves as you want new plants. The new plantlets appear where stem and leaf coincide. Each leaf produces only one plantlet, so this is not as productive a method as midrib and lateral vein cuttings, but you can often purloin several leaves

from each plant without doing it any disservice. When the tray or pot is labelled, water thoroughly with a fine rose, which will settle the cuttings in. Several plants can be propagated from leaf petioles, including peperomias, crassulas, and begonias with smooth leaves.

ROOT CUTTINGS

Taking cuttings from a plant's roots seems an unlikely thing to do – dividing it with roots, shoots, stems and leaves intact is an obvious way to make more, but using just the roots on their own? How can that work? Even experienced gardeners shy away from growing new plants from root cuttings. Chances are, if someone in the family has passed on their skill in propagating plants, their forté will be taking stem cuttings or growing from seed.

Expertise in taking root cuttings has generally remained within the domain of professional gardeners and, like other more 'specialized' techniques, is surrounded with an aura of mystery. Unforeseen difficulties are hinted at, and many of us are inclined to leave well enough alone and not to trespass into the areas of the expert or the specialist. Yet just like most methods of propagation, the practice of growing new plants from roots is very simple.

The hardest part is identifying which subjects can be successfully increased in this way. In every substantial gardening textbook there are lists of plants that lend themselves to the technique, but as a starting point our own personal observations can give us a fair idea. All propagation is based on emulating nature. Observing what happens in our gardens and in the wild gives us clues about which plants to use and how to transform their roots into viable plants.

For example, who hasn't tried to move an oriental poppy (*Papaver orientalis*), digging down to a great depth and making a really thorough job of it, only to find it reappearing the following spring with even more vigour and determination? It doesn't take any expert knowledge to realize that these opulent beauties need no trace of leaf or crown to regenerate *ad infinitum*. They make perfect candidates for this method of reproduction. There is a big double bed (double flower bed, before you get the wrong impression) at Glebe Cottage that we call the hot bed, which bisects the light, bright, open side of the garden. Since our garden is terraced, the hot bed's north edge is level with the brick garden, but its south edge has a retaining wall about 1.2m (4ft) high made of sleepers to make the whole bed level, so the soil is deep. A number of plants of *Papaver orientale* (Goliath Group) 'Beauty of Livermere' were planted through the border to make bold splashes of red right at the start of summer. All had been propagated from root cuttings, and there was great excitement when we put them out. Well, we all know what happens to the best laid plans: by some mix-up one was pink, and since most of these oriental poppies look similar at the leafy stage, the intruder wasn't discovered until it started to flower. Beautiful flowers, too, of pretty pale coral, with frilly edges to boot, which really rubbed it in. After several attempts to remove it, we decided that it was just meant to be there. A bronze-leaved elder that lives nearby has flowers of the very same colour, so we have submitted to the garden's will. Since it seems so persistent, so determined, a few of its roots are borrowed each year to make more plants – but nowadays they are clearly labelled and are planted out where we decide they look best. Though come to think of it...

Similarly, acanthus are almost impossible to eradicate once they have made themselves at home; attempts to remove them usually result in a forest of stems replacing each one that was carefully dug out. In Alice's garden, the next bed down from the hot bed, a stray acanthus took root. It presumably started from a seed, because there are acanthus in other parts of the garden. Its spiky leaves looked so handsome, so glossy, that it was left – and of course it promptly decided to make itself far too comfortable. My friend Danny was working here at that time, and his speciality was digging holes. He had even done some grave-digging, and he seemed just the person to unseat the acanthus when it got too big for its boots. I had to go out that day, so explained to Danny that every trace of acanthus root must be removed and that, though it hadn't been growing there long, they might have travelled to quite some depth. When I came back

several hours later Danny was still working there, but only his top half was visible: he was standing in the hole he had excavated and still hadn't got to the bottom of the roots.

'Japanese' anemones, cultivars of *Anemone hupehensis* or *A.* x *hybrida*, are notoriously prolific once they become established, so much so that when they have completely overrun a garden, bundles of their roots are often proffered as gifts to those just starting or moving into a bigger garden. Perhaps it's just as well that these donations are usually great woody chunks that seldom settle down. The young, slender roots stand the best chance of establishing themselves. However much they spread, colonies of the beautiful pink *A.* x *hybrida* and the elegant *A.* 'Honorine Jobert', with its exquisite white petals, make one of the most uplifting of autumn's pictures – not just in the shady reaches of the garden at Glebe Cottage, but wherever they gain a foothold in the front gardens of city streets or suburban back gardens. In nature, the Japanese anemone is a woodlander, often growing in thin soil over rocks. Its stoloniferous roots spread themselves far and wide, gaining a foothold wherever they can. All along those roots are a series of tiny, embryonic leaf nodes, each one of them capable of making a new shoot and eventually a separate plant. This ability to proliferate spontaneously from a creeping root is extremely useful when conditions are hostile.

Only a select group will grow from root cuttings: the European eryngiums with their spikey bracts make ideal candidates, but the South American types, like *Eryngium agavifolium,* just

ORIENTAL POPPIES
Brief shooting stars they may be, but when they are in full spate nothing can compare with the huge, tissue-paper blooms of these blood-red poppies. You can have as many as you like from root cuttings.

rot away. (See pp.216–19 to match different plants to suitable techniques.) The plants that lend themselves to increase by root cuttings have often evolved a survival strategy that enables them to come back fighting when their top growth is damaged or completely destroyed. Even when these plants have been grazed, trampled or subjected to disruption from the erosion of soil, they can still hold out to grow again another day.

Many of these are herbaceous plants that look best in groups or as a recurring theme in a border. When planting for dramatic effect, it is exciting to have a large number of the same plant for maximum impact and thrilling to think you grew them all from a few pieces of root.

Choosing and collecting material

Taking material for root cuttings of any kind couldn't be simpler.

Gently delve into the soil on one side of an established plant until you come across young roots, cut and carefully extract them – or lift the whole clump, shaking off some of the earth and severing the best roots close to the crown of the plant.

Alternatively, if you want to propagate a new variety, buy a good, healthy, well-established plant and when you get it home tip it out of the pot to expose the roots. A well-grown plant will have several roots running around the outside of the compost, which can be detached without doing the parent plant any harm; often, nascent shoots will already have started to grow along the root.

One very organized school of thought suggests digging up prospective donor plants the season before material is due to be taken, cutting off most of their roots and replanting them. Just as pruning a shrub will stimulate vigorous new growth, pruning back old roots will also encourage the formation of strong, new roots, which are ideal for root cuttings. But if you dig up a plant to prune its roots, it is difficult to resist using those prunings for new cuttings, especially if they seem robust, so if I followed this practice I might end up with more cuttings than I needed. It's always difficult to throw anything away when you know there is a chance it will make new plants. Should you dig up a plant that has been in the garden

for a few years, there will be roots of different ages and most of them will make viable root cuttings.

Not all roots are treated the same way: if you want them to work, you have to do what nature does. There are a few plants, some are woodlanders, that send their roots creeping along close to the surface, ready to pop up new plants along the way. These are fed first from the parent plant and then by the roots along the way, so they can travel until conditions are right and then really get going on their own. The growth starts from little nodules, which you can feel if you run your fingers along the roots; keep these roots horizontal as cuttings and they will reward you by producing new growth from all the nodules. But if a plant has roots that grow down, as most of them do, keep your cuttings vertical, even if you can feel nodules along them. They will only produce new roots from the bottom and shoots from the top.

TAKING TIME

Although taking root cuttings is a long-established method of propagation, very little research has been done on the when of it. Pundits now think that the optimum time to take root cuttings is in the middle of the plant's dormant season. With most herbaceous plants this would be around the turn of the year, but some spring flowering plants, such as *Pulsatilla vulgaris*, start growth early in the year so would be at the depth of their dormancy in mid-autumn. At Glebe Cottage we take root cuttings from autumn through to spring, and most are successful.

IDEAL MATERIAL
The roots of some plants are constantly questing for new ground – even pots cannot constrain their colonising instincts.

All our root cuttings are kept on heated benches. A small heated propagator will take several pots of cuttings, and can be used throughout the rest of the year to promote rapid rooting of stem cuttings and germination of seed. It will never be empty and soon pays for itself. In fact you may very quickly find yourself investing in another one, or a bigger one, or both. Even if you have no bottom heat (always a sad prospect!) cuttings placed in a coldframe or on a greenhouse shelf or warm, bright windowsill will soon root.

Vertical cuttings

Cuttings from the roots of some herbaceous plants are irrepressible, and seem to turn into new plants automatically.

Oriental poppies are a typical example: anyone who has tried to move one will know that the bits that are left behind often make the best plants. Exploit this by taking a few of the strongest roots and making new plants. New cultivars abound, and there is everything from the big and blowsy 'Picotee' and 'Curlilocks', to the neat and pretty 'Karine' and 'Juliane'.

Lift or dig around a plant – or decant a potted one – to expose a few thick, young roots, and simply detach them. Chopping off a few chunks will not affect a healthy plant when you come to put it into the ground. The most important thing with vertical cuttings is to keep them the right way up. Polarity is all important, because the cuttings won't grow if they're upside down. Books often advise making two cuts, straight across at the top of the cutting (closest to the crown of the plant) and at an angle at the bottom, solely to distinguish one from the other, but providing the cuttings are lined up alongside one another and promptly pushed into compost this is unecessary. Cuttings should be 2.5–5cm (1–2in) long, depending on their thickness: the skinnier the cutting, the longer it needs to be.

Push the cutting into the compost so that its top is flush with the surface. If the roots are on the thin side, use a

ACANTHUS ROOTS
OPPOSITE These cuttings of the cultivar 'Hollard's Gold' were taken straight from the pot and lined up so they did not end up bottoms up. Buried to the hilt in modules, topped with gravel and watered in, they made plants ready for the border in a few months (overleaf).

chopstick or dibber to ease them into the compost. Cover with a layer of grit and water well. Some gardeners advocate using fungicide, but this is neither necessary nor desirable; if material is healthy, the compost well drained, and the cuttings well aerated and kept in a warm, bright place, there should be no problems.

The first sign that things are going according to plan is the appearance of new leaves on the top, cut surface of the root. New leaves develop before fibrous roots, so wait for two or three weeks after leaves emerge before potting up the cuttings into individual pots, checking that a new root system has developed. Grow on in the greenhouse (a heated propagator is a boon) or a coldframe and plant out in permanent positions when well established. Give potted cuttings several months to establish before planting them out.

OTHER SUITABLE CANDIDATES
All species and selections of acanthus will multiply from root cuttings, and material is usually easy to find, because acanthus roots are thick and fleshy. As well as the golden-leaved forms of *Acanthus mollis* – 'Hollard's Gold' and 'Fielding Gold' – there are some striking and unusual species to try. If you don't have them in the garden, buy a pot of *A. spinosus* ('Spinosissimus Group') or *A. hirsutus*, and you should find that there are masses of roots that you can plunder.

113

Not all pulmonarias come from root cuttings, many are best divided (see p.176), but forms of *Pulmonaria longifolia* and cultivars that have this species in their genetic make-up are easily propagated in this way. We have had success with *P. longifolia* 'Ankum', a lovely silver-leaved form with purple flowers, *P. longifolia* 'Dordogne', with bright blue flowers, and *P.* 'Moonstone', a tall, spotted-leaved form with opalescent flowers – a sport from our own plant *P.* 'Glebe Blue'. Slice a few thong-like roots into chunks about 2.5cm (1in) long. Fill a clay pot or module tray with gritty compost and use a dibber or a chopstick to make holes around the edge of the pot or in the centre of each cell. Push the cutting in until it is flush with the compost surface. Finish off with sharp grit, water once, and place the pot or tray in the warmest place available. A heated propagator is ideal, but even in a coldframe the cuttings will succeed, they'll just take a little longer. The severed roots will produce a rosette of leaves and shortly afterwards new roots will form. Don't be in too much of a hurry to move them on and when you do, be very gentle – this is why I prefer to use modules, since the new root system is not disturbed when potting on. When plants are big enough they can be planted out into the garden in groups of three or five. What luxury – and all for nothing!

One of the most vibrant blues of all in the garden, and very visible thanks to its 75cm (30in) flower stems, is *Anchusa* 'Loddon Royalist', but it only has a short life and will not come true from seed – in fact I can't remember ever having seen it seed set – so the traditional way to increase it is from

root cuttings. There are other worthy anchusas too: if you prefer paler blue, 'Opal' is outstanding. Comfrey (*Symphytum*), a close cousin of anchusa, can be propagated in the same way. If you want to grow a lot of leaves for the compost heap or to make liquid comfrey, or you just have a plant with particularly deep-coloured flowers, then root cuttings will produce plenty of plants relatively quickly. Taking root cuttings is the best way to increase both *Crambe cordifolia*, native to the harsh landscape of the Caucasus, and *C. maritima*, sea kale. Seed may ripen well on crambes in a good year, but their tap roots are perfect for root cuttings, and this usually is the best way to make new plants.

Primula denticulata is an easy plant to grow from seed, but if you want to make clones of a plant with a distinctive flower colour or an especially neat form, apart from sending it to the lab for tissue culture, root cuttings are the only way to go.

FORCING ROOTS IN THE GROUND

There is another way to persuade some shrubs growing on their own roots to yield new plants. Dig around the perimeter of a suitable shrub when its leaves have fallen, using a sharp spade to cut right through the roots. It should be a fairly young specimen and if you have a planting hole ready elsewhere, lift it and replant – it should settle down quickly. It can be put back in its original space, but that might overshadow the roots left behind and inhibit them from producing new shoots, which is what you are after.

Eventually, the severed roots that are left in the ground will form new shoots, just like the roots that are inadvertently left behind when digging out perennials will, and each one should make a viable plant. The time this takes may vary from species to species, but when new shoots are visible, preferably during the next dormant season, each one can be lifted and replanted or potted up.

LIFTING SUCKERS

Several shrubs can be increased from roots or rooted 'suckers' that sprout along their roots. Raspberries, lilacs (*Syringa*), and the shimmering *Elaeagnus* 'Quicksilver' all spread in this way. You could say that these plants spontaneously make their own root cuttings.

We have a lovely crimson-leaved form of *Prunus padus* called 'Colorata', which spontaneously makes new shoots feet away from the main trunk.

If we need more (they make a great addition to our native hedge), the root attaching them to the main trunk can be severed (cut through with sharp secateurs), then during the dormant season the whole shoot, together with its roots, can be carefully lifted and replanted. They could be potted up instead, if we wanted to give them away or plant them out at a later date. The idea of severing the roots, but

waiting until later to lift it, is just to ensure that the new plant goes through as little trauma as possible.

In the same border is *Rosa glauca*, a species rose with grey foliage, single pink flowers and red hips that eventually turn jet-black. This vies with the cherry, suckering here and there and giving us new plants. Since several of the hybrid roses in the garden here are growing on their own roots, propagated

initially from layers or cuttings, any errant shoots they may throw out can be treated in the same way. This is not the case for suckers thrown up around roses grafted on to a rootstock (most of the roses you can buy) nor, come to that, for any other plants grown on rootstocks. The sucker will always be the same as the rootstock, never the variety of rose, apple or whatever else is grafted onto it.

Crambe The tap roots of crambe are fleshy and thick

Crambe cordifolia peaks at an impressive 2m (6ft). It takes some years to reach this stature, but in fertile soil new plants get off to a good start and often produce some flowers in their second year. It is an odd-looking creature in the depths of winter, a huddle of strange, sleeping shoots, each one terminating in an almost comical purple bud. This is the best time to take root cuttings.

1

2

3

4

5

– perfect for making new plants and quicker than seed.

6

7

8

9

10

ROOT TO PLANT

1 Don't lift the plant, just dig down deep to detach roots.

2 Sever roots of about pencil thickness from the plant.

3 Cut the roots into chunks about 2.5cm (1in) long.

4 Insert individual pieces into cell or module trays in loam-based compost with grit.

5 A generous layer of grit keeps down liverworts and mosses and retains moisture.

6 After a few weeks in a bright spot a flurry of leaves emerge.

7 It's not long before the new leaves develop.

8 Plantlets can be potted on when proper new roots have formed.

9 Each one has its own pot to encourage it to grow on strongly.

10 Finish off the pots with grit and put in a sheltered place to grow on.

Horizontal cuttings

All root cuttings are easy to take, but horizontal ones could not be more simple – they don't even need to be buried.

All the Japanese anemones can be increased by this means, and root cuttings can be taken from them successfully at almost any time of the year. We do it here in early to mid-autumn; the appearance of all those glorious pink and white flowers prompts us to get cracking. This is the most convenient time for us, but others do it at different times, and it usually works for all of us.

Whether you are exposing the roots of a plant in the ground or lifting a new plant out of its pot, the roots you are looking for should be young, pale brown, and slender, and about the thickness of darning wool. Trim off any hair-like roots along them and slice the main root into pieces. The length of these pieces is not crucial; if you are going to grow the cuttings in modules, simply cut them to fit across the compartments. Growing one root cutting per module or cell is the best method, especially when it comes to potting up. Each cutting will make its own root system, which will suffer the minimum of root disturbance, so new plants establish fast with no check to their growth.

Use a gritty mix and include either sterilized loam or John Innes compost to keep the cuttings going for a while before they are potted on. Lay the cuttings on the surface of the compost

and press them down firmly, keeping them horizontal and making sure that they are in intimate contact with the soil. Weight them down with a covering of grit, which will also help to retain moisture and provide sharp drainage around the developing shoots.

A thorough watering from above should ensure they are nestled in happily, but always use a can with a

SKINNY ROOTS
Sometimes, when you decant a Japanese anemone the roots are already making new shoots. A plant can yield scores of root cuttings. Our *A. x hybrida* 'Snow Queen' (opposite) was a sport, now propagated this way.

fine rose for this, to ensure that nothing gets washed away. Most of the cuttings will develop several new shoots along the surface of the root, creating a bushy plant from the start.

OTHER SUITABLE CANDIDATES

Cranesbills, or hardy geraniums, are among the most popular perennials, and rightly so since they are such attractive and accommodating plants; there are geraniums for every kind of soil and situation. Most species will come easily from seed, and some cultivars can be increased by careful division or even by taking basal stem cuttings first thing in the spring. But one very popular species, *Geranium sanguineum* and its cultivars, can be increased most easily by root cuttings in exactly the same way as the Japanese anemones. We grow several cultivars of this sun-loving geranium, and one particular favourite of ours is *G. sanguineum* 'Ankum's Pride', which was selected by a brilliant Dutch nurseryman, Coen Jansen. Its flowers are vivid pink, with none of the blue-pink that is characteristic of most *G. sanguineum*.

When the knobbly roots are exposed it is immediately clear that the plant has the potential to make roots from any of the nodes along the length; sometimes embryonic new shoots can already be seen. The wild species can cover a large area simply by pushing up these shoots from its expansive root system, and it is this propensity that we exploit when taking root cuttings. Dig around a clump of the geranium using a small hand fork to expose the roots. Those on the outside of the clump will be the newest, chunkiest and most vigorous, so they make ideal

material for cuttings. The procedure is just the same as it is for propagating Japanese anemones, but since the geranium roots are a little thicker, they should be pushed slightly deeper into the surface of the compost. Just as with anemones, roots for cuttings can be harvested from a plant in a pot as well as in the ground; in potted plants, the youngest, most thrusting roots will have forced their way to the edge of the rootball, against the pot wall.

NEW ANEMONES
Short pieces of root are laid horizontally on the surface of gritty compost and weighted down with grit. In a matter of weeks, new shoots spring up, and a few months later the young plants you've potted on are ready for the garden.

The big picture

You could hardly guess that this elegant plant, with its clouds of honey-scented white flowers is a brassica; only when its clump of large, dark green leaves is in view would you suspect the cabbagey ancestry of *Crambe cordifolia*. Although the flower stems are stiffly held, there is a lightness about the plant that lifts the whole border and makes it twinkle – espccially at dusk. This is the same plant in its summery glory from which I'm taking root cuttings (page 114) in the depths of winter.

BULBS, TUBERS AND RHIZOMES

BEAUTIFUL BULBS
Bulbs produce some of the most gorgeous and sumptuous flowers on the planet – as well as some of the most exciting and outlandish. Increasing them is an adventure, and there is much to learn about the plants by propagating them.

Soon after we first came to Glebe Cottage, I heard of a gardener called John Huxtable who lived nearby. He was a friend of a group of keen plants people who gardened in and around Porlock. Norman Hadden called him 'Farmer John off the Moor'. Sadly, I never got the chance to meet him, but I did see his garden. Even with no-one to tend it for months after he died, it had a magical quality, sparkling with jewel-like flowers and inhabited by rare trees, climbers and shrubs. It was one of the most sublime gardens I have ever had the honour to experience. A year or so later I was invited to visit again to see if there was anything I would like cuttings from. Climbers had been removed from the walls of the cottage, and the garden was devastated, but just across the way was a little fenced off plot, obviously used for many years by John Huxtable as a nursery garden. In the grass were old lead labels with the legends of lily species and other bulbs recorded on them. Some would have been grown from seed, some from offsets, but more striking evidence of the plot's history were the stems of *Lilium svotzianum*, still rising from the grass and brambles. Bulbs are the most tenacious and rewarding of plants, and propagating more of them is a joy.

Though a few gardeners have promoted naturalizing bulbs (two of the most influential being William Robinson in the late 19th century and Christopher Lloyd in our own time), it is only in recent years that bulbs have begun to play a major role in informal planting schemes on a popular scale in smaller gardens. Formal bedding schemes and pots used to be the two most important employments for bulbs, but both these situations require uniformity, and almost always bulbs were bought afresh each year. At Glebe Cottage we buy tulips every year for pot displays, but in other parts of the garden, snowdrops (*Galanthus*), narcissi, camassias, erythroniums and trilliums are left to their own devices. Elsewhere, we enjoy the glamour of irises, the opulence of peonies, and the downright worldliness of our favourite potatoes – all of which we increase ourselves.

Nature has devised a multitude of ways for plants to cope with tough environments. What we group together as bulbous plants technically fall into several different categories, apart from the true bulbs with their packed layers of modified leaves. Where a plant faces cold winters and scorching summers it may have developed swollen root tubers to store food over the winter, supporting leaves and stems that

expand rapidly in the growing season to produce abundant flower and seed, but collapse at the first frost. Several familiar plants have developed this way, including dahlias and sweet potatoes. Another adaptation is seen in plants like potatoes, which have stolons (underground stems) bearing thickened tubers to store food reserves. In other plants, such as bearded irises and hedychiums, stems themselves have been modified to become storage organs, either squat, vertical corms or horizontal rhizomes. The plants in each of these groups all have their own idiosyncracies, and propagating them is about what works, not what the rules say should work.

Plants that grow from true bulbs and corms often increase themselves – bulbs divide spontaneously to make smaller bulbs and corms make new corms on top of the existing ones. It is not immediately obvious where and how human intervention could replicate or speed up this process. There is no branching stem that can be commandeered for cuttings, or tap roots that could be chopped up and used as root cuttings.

ABOUT TO PERFORM
The tulips have gone, now it's the lilies' turn. Lilies add a touch of panache to any border. They are often used most effectively as a moveable feast, grown in pots and brought in to liven things up. It's impossible to have too many.

True species can sometimes be raised from seed, but bulbs are notoriously slow. There are exceptions like bluebells (*Hyacinthoides non-scripta*) or the dreaded *Allium triquetrum,* which increases at sci-fi speed. But in many cases it can take years for seed to turn itself into flowering bulbs. Trilliums, members of the lily family, may take as long as seven years, but there is no other satisfactory method to increase them. Other genera produce mature plants far more quickly from seed, but often the most special of them are hybrids that will not seed true. So if you want to increase a special plant as rapidly as possible, are there any simple methods of propagation that can produce enough to make a show?

When bulbs or corms are happy they will increase; we can make the most of their colonizing habits, simply digging them up and spreading them out. But if your plants lack world-conquering colonizing tendencies, you can steel yourself and cut them up, provoking them to make 'mini-me' bulblets.

It is difficult to lay down hard and fast rules about how to increase tubers and rhizomes, and every one needs to be considered individually. Some tubers are indivisible and can only realistically be increased from seed. Cyclamen are a good example; their tubers can reach magnificent sizes, and each may mother enormous numbers of progeny through seed, but each will never become anything but an individual. Some tubers ignore the rules and behave like bulbs, like *Arum italicum*; *Zantedeschia aethiopica,* another aroid, is a rhizome, and both of these can be multiplied from their roots, as well as from seed. Since this is a kind of vegetative propagation, the

new plants will be carbon copies, clones of the parent plant. In many cases there may be other ways of vegetatively propagating the plant – for instance growing dahlias from basal stem cuttings (see pp.80–5) – but it may be more convenient to make more by splitting the plant and breaking up the tubers or chopping up the rhizomes. It is instantly gratifying too, because you can see that your new divisions have everything they need to grow into established plants very quickly.

When a plant gives a breathtaking display year after year, there is a great temptation to leave well enough alone. We expect clumps of flag irises to go on producing glorious velvet-and-satin flowers without fail. We assume the crocosmias that made their debut to gasps of delight will carry on in the same vein *ad infinitum*. But in the end, flowers will dwindle and leaves become more sparse. We may be brave enough with clumps of perennials that have become bare in the middle – they seem almost indestructible – but revitalizing these plants by division seems fraught with problems. When do you do it? How much of the original rhizome do you need? How do you replant? Sometimes it seems better to ignore the whole thing and leave it to its own devices.

Each type may have its own esoteric nature, but broad generalizations do apply in each case. To make new plants takes three things: a food source, a new shoot (however embryonic), and new roots or a means of making them. The only other requirements are some common sense, the ability to observe, and the confidence to go for it. It's always worth giving your favourite bulbous plants the chop – and it won't cost you a penny!

Making new corms

Corms have a different construction to that of tunicate bulbs, and increasing them could not be simpler.

The corms of plants like crocosmia, crocus and gladioli are really swollen stems that act as storage organs. On close inspection the embryonic buds can be seen and felt all over the surface. Corms are solid and have no layers, as is the case with bulbs like those of hyacinths and narcissi – or think of an onion. Corms tend to build a new corm on top of the old one each year. The old corm usually shrivels away beneath its replacement – the one really well-known exception being crocosmias, which increase themselves steadily by building chains of corms that can be divided (see pp146–7). But for most corms, or if you have a special variety that you need to bulk up quickly, this is how to do it.

CUTTING UP YOUR CORMS

In the early autumn, healthy corms can be cut into chunks, making sure that each has at least one embryonic bud. Most books advise dusting the cut surface with a fungicide, but as an organic gardener I'd argue that this is unecessary. If the pieces are placed on a wire tray and kept in a warm, dry place, the cut surfaces should heal quickly. The airing cupboard is ideal for this; it only takes a couple of days, so no-one will run out of sheets or towels before the chunks are taken out and potted up or planted out. A corky layer will have formed over the surface and, given sharp drainage, the pieces should not rot. They will form leaves and eventually become a new corm.

CROCOSMIA CORMS
Nothing could be more straightforward than increasing crocosmia corms – they are simply sliced through and sliced again. Each quarter, with its own share of crown and root and one embryonic bud, should be given its own pot to make a new corm in. For run-of-the-mill crocosmias this might seem a real fandango, but for choice cultivars it is a simple and free way to make more precious plants.

Corms that are too small for the chop can be planted out into a nursery bed, after first removing the old stems and roots. Here, they will build up, making several new shoots and turning into decent-sized corms that are ready to make more the next year.

DETACHING CORMELS

Corms often produce cormels, a series of tiny new corms between the old corm and the new. This phenomenon is typical of gladioli, which can produce as many as 50 cormels on each corm. They are formed towards the end of the summer and, when corms are lifted for their winter sojourn in the greenhouse, can be detached and stored in a cool, dry place. Because they are frost tender, storage temperatures should never drop below freezing, but most people have handy corners underneath the greenhouse bench, or in the shed or the garage, where they can spend the winter in boxes or seed trays in dry soil or compost.

Sow them the following spring, either in trays or in the open ground, just below the surface of the soil or compost, making sure they are well marked. They may need soaking in warm water before they are sown to make sure they are plump as can be, especially if they have become a bit dessicated. Results will not be instantaneous, but after a couple of years, corms should have reached flowering size. If they are planted outside, they will need protection during subsequent winters. A cloche, a piece of insulating material, or extra soil built into a ridge down the row should give the corms the protection they need. The time to do this is as the foliage dies down and temperatures begin to drop.

Scoring and scooping bulbs

Persuade daffodils, hyacinths, and other bulbs to make tiny bulblets by going in with the knife – or teaspoon!

The best time for scoring and scooping is when the bulbs are dormant, just before they start to put out roots and begin their new cycle of growth. Fresh, plump bulbs from your newly delivered bulb order are ideal candidates, or even fresher bulbs can be lifted from your own garden; as long as they are buxom and haven't yet started to send out their first roots. In the second scenario, it's worth lifting a couple of daffodils from late summer to early autumn to see just what stage they have reached. This method should work even if bulbs are lifted a few weeks before roots emerge, so take a chance and try it again a couple of weeks later too. With luck on your side, both lots will work.

SCORING BULB BASES

Bulbs can be scored with a sharp knife, making two cuts at right angles to each other to a depth of 6mm (¼in) into the base plate of the bulb. Invert the scored bulbs and put them in a warm, dry place until the cuts open up to form a star. Keep them upside down and put them on a wire tray (the sort you use to turn out a cake or loaf) and put them into the airing cupboard or somewhere similar. It will be ten weeks or so before the small bulblets that will be produced on many of the cut surfaces reach a sufficient size to move them. When the bulblets are almost touching, plant the bulb in a pot of compost, still inverted, so the bulblets are just under the surface of the compost. They will grow on, producing roots and leaves,

while the mother bulb shrivels and virtually disappears. At the end of the season, you can detach the new bulbs and plant them in a tray or in the ground in fresh compost; remember to protect them if they are frost-tender.

SCOOPING OUT BASE PLATES

Sharpen one edge of an old teaspoon using a file or grinding stone – it sets your teeth on edge, and you must remember to keep it far away from legitimate teaspoons! Use it to dig into the basal plate of the bulb with a circular motion, so that the whole plate is removed without digging into the heart of the bulb. Just as for scoring, allow the cut surfaces to callus over by placing scooped bulbs upside down in a warm, dry place. They can then be put into the airing cupboard on a wire tray or upside down in a dish of

PROVOKING ACTION
When bulbs are damaged they are sometimes spurred into producing new bulblets. We can initiate bulblet formation with tools like a sharpened teaspoon or an ordinary knife.

dry sand until new bulblets have formed on the cut surfaces. This may monopolize the airing cupboard for a long time, since it can take two or three months. When bulblets have formed, plant the whole bulb upside down in a pot of compost with the bulblets just protruding from the surface. They will produce leaves during the following growing season and when these die down, towards the end of the summer,

the tiny bulbs can be pulled apart and potted individually or planted out in rows. It is crucial that the bulbs do not dry out – their tiny roots are very sensitive at this stage.

Growing bulbs in this way demands patience; it may be years before the new bulbs flower. On the other hand, the eventual yields can be great, and more importantly the satisfaction from growing your own bulbs is enormous.

HEALING TIME
After scoring, placing bulbs on a wire tray allows the cuts to open and callus over, prior to the production of tiny new bulbs.

Garlic Some alliums are grown from seed, but all the differen

Flowering is the goal, the object of the exercise with the propagation of most bulbs, but there are exceptions. Onions and garlic are grown to be eaten – looks are not paramount and we would rather they didn't flower. Garlic is bought as a bulb and separated into individual cloves that can be planted direct or pushed into modules, trays or boxes and transferred into the open ground. With onions, garlics and shallots it is advisable to plant into the top of ridges, especially if you are on clay soil like ours and/or in an area of high rainfall. This both helps the bulbs to ripen by exposing them to the sun and prevents rotting by improving drainage. Many bulbs from hot, sun-baked, poor soils resent rich fare, but not the onion clan. Give them good fertile soil for best results.

1

2

3

4

5

garlics, strong and mild, can be propagated from their cloves.

6

8

7

9

CLOVE TO BULB

1 If your soil is cold and wet, it is a good idea to start garlic cloves in modules.

2 Modules allow plenty of room for cloves to develop individually.

3 In no time at all, each clove will have a healthy, green shoot.

4 A healthy root system indicates cloves are ready to plant out.

5 Don't dig holes for your cloves; instead, plant them on ridges.

6 The raised ridge of soil promotes good drainage around the cloves.

7 Being above soil level also means each clove gets maximum warming from the sun.

8 Some garlic cultivars run to seed easily.

9 Even if your garlic flowers, the outer cloves will remain plump and there is the bonus of tiny new cloves in the seedhead.

Scaling bulbs

Bulbs can be increased by persuading each layer or scale to produce a new bulb at its base.

Lilies and fritillaries have open-scaled bulbs, like garlic, rather than the solid, tunicate bulbs of narcissi. They may produce side bulbs, but scaling is a more efficient means of propagation. The plumper and more turgid the bulb, the better the chance of success. Snap clean, healthy scales from the parent bulb, as close to the base plate as possible. This is most easily achieved on new bulbs received in the autumn or spring, but it is entirely possible to snap off scales from plants growing in the garden, by carefully removing soil from around the bulbs and treating them in the same way.

Make up a mixture of peat, or peat substitute, and grit or pure vermiculite and mix in the scales. Half-fill a polythene bag with the mixture. Blow into it and tie it up tightly, trapping air inside. Store in the airing cupboard (it's getting quite full, isn't it!) and check from time to time. As soon as new bulblets start to form, which can take a couple of months, pot them up with the scale still attached and the bulblets just peeping out from the compost surface. The bulblets will still draw food from the scale, even though they are beginning to produce their own root system. Cover the surface of the compost with grit and stand on the greenhouse staging or a bright warm windowsill. The baby bulbs will make leaves during the summer and later on, after the leaves have died down, the bulbs can be separated from the now-withered scale and planted out.

Because these bulblets will be more substantial than those that are made by some other methods, and because in most cases they are completely hardy, they are better able to fend for themselves. They can be planted in little rows in a flower bed, the vegetable garden or, better still, in a bespoke nursery bed or soil-filled coldframe. If this sounds far too grand, it needn't be: it can be as large or small as you have room for, tailor-made to your own needs and space.

TWIN-SCALING

More tightly-packed bulbs that have a basal plate, such as some members of the lily and hyacinth families as well as narcissi and snowdrops, can be propagated by twin-scaling. This isn't something I do, but specialist bulb nurseries use it to produce the large quantities of bulbs they need each year. If you have a special bulb that doesn't naturally produce offsets very often, or you really want to bulk up your numbers, this is the way to do it, because you can get a couple of dozen bulbs from just one (the really skilled professionals can get several dozen). Most will take some years to flower, although snowdrops can be speedier.

Like scoring and scooping, this is a way of forcing a bulb to produce little bulblets by damaging it – pretty severely – and like those methods, it needs a healthy, dormant bulb and clean, sharp tools. First the bulb is sliced top to bottom into anything from

LOTS OF LILIES
One of the most graphic ways of propagating is growing new bulbs from lily scales. A few scales can be removed with no danger of spoiling the performance of the mother bulb, and if they are provided with the right growing medium and a warm, dark place to get on with the job, new bulbs are almost magically produced within just a couple of months.

quarters to sixteenths, depending on its size. Treated like the scales of a lily bulb, these 'chips' will produce bulblets from the basal plate, nestled inside the outermost scales, in about 12 weeks. Twin-scaling takes this a bit further, dividing each chip into pairs of scales, each with a bit of the all-important basal plate attached: each pair will produce bulblets at the base between them, increasing the yield. When the bulblets appear, plant the chips or twin-scales individually, but intact, with the bulblets under just 1cm (½in) of compost and the scales exposed.

Increasing your own bulbs is one of the easiest and most fulfilling ways of propagating, though it does demand more patience than some other methods, especially in terms of the length of time you have to wait to see your new plants in flower.

Lifting and separating

Naturalized bulbs are often left to their own devices in situ, but they will flower better and spread faster if you give them a helping hand.

This is done either during their dormant season or, in the case of snowdrops and some others, soon after flowering, when their foliage fizzles.

The colonizing characteristic of these bulbs can be exploited by the gardener to increase their stock of bulbs and achieve show-stopping effects. The best way to see snowdrops is carpeting a woodland floor in huge drifts, as far as the eye can see. Few of us can emulate this sort of picture, but most of us can enjoy snowdrops planted amongst our perennials, shrubs and trees, thickly enough to create the same uplifting effect.

The better you grow your snowdrops, the faster they will increase. That is not to say they should be given a rich diet, just one tailored to their needs. Since most are woodlanders, give them what they would get in nature, which is leaf mould and woodsy soil – well rotted, home-made compost is an ideal substitute. When you are dividing and replanting, sink them into the same comfortable mixture, water well and let them make themselves at home.

The ideal time to divide snowdrops is just as their flower stems fade. If you have species or selections that set seed, you may want the seed heads to ripen and seed to fall before you move them, though if they are replanted promptly

they should distribute their seed in their new position. Sometimes you just don't have time to do things when you ought, but if you missed the boat as foliage was dying down, don't despair. It is bad news to plant snowdrops as dry bulbs – that is, in the same state you would daffodils or tulips bought from a bulb merchant or garden centre – because dry snowdrops will often remain in a state of dormancy, and sometimes never emerge. Such bulbs may have been stored for months, but your own snowdrops in the garden, even if you haven't seen them since the spring, will still be very much alive.

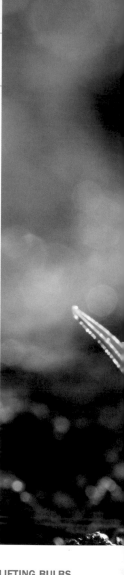

LIFTING BULBS
There are basically two schools of thought on when to divide snowdrops. One decrees that they must be lifted and split 'in the green', the other says they can be divided while they are dormant, but not dry.

RAPID REPLANTING
If you are prompt and replant bulbs singly after enriching the soil, you are guaranteed possibly the finest, and certainly the most iconic, sight of spring.

Providing you can still locate them, you can dig them up and replant them successfully later in the year, with no trepidation about their flowering next spring – the key requirement is that they are replanted immediately after they have been lifted and separated. The advantage of being a bit late is that new bulbs will have been formed on the side of very mature bulbs, and these can be detached and planted separately somewhere else or included in the new scheme.

When replanting, whether it be at foliage-fading time or later, avoid shoving clumps of bulbs into one hole and repeating the process through the whole scheme. It makes my blood boil when I see items on television where a presenter digs a series of holes and plonks half-a-dozen or so bulbs into each of them, usually unseparated and just peeled off a bigger clump. The whole idea of dividing up these bulbs is to give them the best chance. Not only is it good for them – and surely we owe it to all our plants to nurture them – but the show of flowers and healthy foliage that they will produce is recompense for any extra trouble or time it may take. If you vary the distances between bulbs, randomly planting some fairly close together, others farther apart, you can create

a naturalistic look. Plant snowdrops and narcissi deep: their roots have the ability to adjust the level of the bulb by contracting and pulling or pushing it to the right level.

Leucojum, both *L. aestivum*, the so-called summer snowflake and its relatively tiny counterpart *L. vernum*, the spring snowflake, can be divided in the same way as snowdrops, as can other tunicate bulbs. Adjust the time of the operation based on the bulb's progress: common sense dictates that the later the bulb flowers, the later we can leave division and replanting.

TREAT AROIDS LIKE BULBS

The aroids *Zantedeschia aethiopica* and *Arum italicum* are very similar and can be propagated in much the same way. You can grow both from seed too, but if you are impatient or enjoy this way of making more plants, dividing is straightforward and fairly foolproof.

Zantedeschia is an evergreen, but in our climate it almost always behaves like a herbaceous perennial, dying down during winter. It is best split when in active growth, but before its shoots are too advanced (certainly before it thinks of sending up its huge white spathes) so that it will have a chance to settle in and make new roots before it has too much to support.

SPLITTING ARUMS
Zantedeschias, and many other aroids, make a cluster of stems with bulbous bases that are easily pulled apart and potted up individually. Some will be large enough to flower the next summer, others may take a little longer.

Clumps are made up of a collection of rhizomes, each with their own roots, and it is a simple exercise to wash them and pull them apart, potting the biggest into individual pots and perhaps planting the really teeny ones in cell trays or modules. Some plants dislike being overpotted, but aroids are not shy – they will quickly fill pots with their fleshy white roots. The best time to make your move is probably in the early part of the year, but if you don't get round to it until later it won't hurt, although your young plants will need a frost-free position in a greenhouse or cold frame to carry on growing.

The glossy, marbled foliage of *Arum italicum* is one of the most welcome sights in the autumn garden. Following on from glorious, bead-like berries, which are first green then rich orange and red, the leaves expand and grow right through to the end of spring, seeming to glory in the cold weather. The berries are consumed by birds (slugs and snails like them too) and their seeds are deposited around the garden. Very often they are in the perfect place, but occasionally you can think of better sites. We had a large group in the middle of a raised veg bed, and redistributed it to a shady spot that needed some winter pizzazz. The procedure is the same as for zantedeschia, except you can carry it out during autumn and winter, anytime that you can see the leaves, and it is not essential to pot up and protect; pieces can be planted immediately into the open ground. Add leaf mould or home-made compost and plant really deep, soaking the newly separated divisions. Try to carry this out on mild days, and never attempt it if the ground is frosty.

BULBS THAT MAKE GRASS AND RICE

Some bulbs make masses of new bulbs without intervention. This can be annoying for gardeners, especially when the bulbs split into so many tiny bulblets they are no longer capable of producing flowers. Cultivars of *Iris reticulata* and *Iris histrioides* are notorious for this, and may come up as 'grass' in their second and subsequent years, there being lots of tiny bulbs and no bulb of flowering size. If these small bulbs are separated and allowed to grow on, they make flowering bulbs much faster than when left to their own devices. One of the most sought-after of these striking bulbs is Iris 'Katharine Hodgkin', with ice-blue flowers finely veined in deeper blue. It tends to disintegrate more slowly than many of its cousins, and if smaller bulbs are detached carefully, the mother bulb will draw in more nutrients from the surrounding soil and is more likely to make goodly flowers, while the infants can be lined out in trays leaving enough space between them for future root development. Some fritillaries make tiny bulbs called 'rice' in large quantities around the base of their stems. These can also be collected and grown on in just the same way. Sometimes pots of rice are sold at Alpine Garden Society meetings, and this can be a good way to obtain stock of some of the rarer species.

Dividing rhizomes

From Monet's Garden to modern Chelsea Flower Shows, bearded irises have won hearts and turned heads.

Their spectacular flowers are addictive, and for decades breeders have given us ever-more voluptuous, eye-catching blooms. Once the gardener is bitten by the bug, that's it.

There are only a few places at Glebe Cottage where we can grow bearded irises: the two beds in front of the house, our little 'seaside' garden and a big raised bed. Conditions are similar in all four sites – full sun, excellent drainage and fairly poor soil. Sometimes we make the soil even more to their liking by incorporating lime rubble, gleaned when old walls are pulled down. The same stuff works magic for pinks too, if your soil is on the acid side: as well as increasing the alkalinity it also gives the crunch that such plants enjoy, a reminder of the soils of their wild habitats. Bearded irises need very little in the way of nutrition; one of their close relations is *Iris* 'Florentina', the orris root that is grown as a catch crop under and between olive trees where it enjoys the poorest, driest and hottest of conditions. It is harvested each year; the spare tubers are dried, and sold to the perfume industry as a fixative.

All irises in this section flower for several years unaided, but eventually they dwindle and give up the ghost. The old rhizomes become dessicated, and the new rhizomes that emanate from them can hardly produce new roots and shoots. In a garden, the rhizomes piggy-back atop one another in a desperate attempt to find pastures new.

When clumps become large and congested, and the rhizomes begin to overlap, dig them up. Knock off surplus soil so you can see what is happening and where the rhizomes are joined. Often there are several side rhizomes growing from one old woody 'mother' rhizome in the centre. Cut or snap off any old, wizened roots: you're after strong chunks with a fan of leaves at one end. Pull outside leaves firmly down so they come off cleanly, and trim the remaining leaves neatly with a pair of scissors or a sharp knife. These irises need sun to ripen them up and produce flowers, because the rhizomes act as a heat-store, so bear their sunbathing requirements in mind when replanting. A double trench 5–8cm (2–3in) deep with a ridge of soil running down the middle is ideal; drape new roots on either side of the rhizome down the sides of the ridge, and back-fill the trenches with soil. The rhizome itself should be just above soil level, where it

GROWING GINGER
Root ginger is a familiar sight on our supermarket shelves. It is a rhizome, and as such it can be brought into growth by the simple act of planting it.

can bake to its heart's content. Plant pieces in groups, directly into the garden, or let them convalesce in a nursery bed to swell and gather their strength for the open garden.

TENDER RHIZOMES

Some plants take all growing season to rev up for their big show. Some of the most ostentatious are from the tropics or sub-tropics, and it shows. There are few spear-carriers here.

The huge heleniums, perennial sunflowers from the prairies of North America, make he-man clumps at the back of the border – perfect alongside imposing stands of filigree fennel. But both are elbowed to one side by the towering forms of more exotic subjects: hedychiums, cannas and bananas (*Musa*). All have broad, paddle-shaped leaves and outrageous flowers. Cannas are often used in municipal plantings, and it can be rather sad to see them used as dot plants in a sea of marigolds or petunias. But a well-placed canna can be a master-stroke: put 'Pretoria' where low sun catches its pink-edged,

dark leaves, and enjoy the rusty orange flowers tumbling from their spike like silk handkerchiefs from a magician's sleeve. The hedychiums, or ginger lilies, are close cousins. Their foliage is green, and their spikes crowded with 20 or 30 flowers, but the big difference between these and cannas is scent. Ginger lilies are pollinated by moths, and exude an overpowering scent at dusk.

Both have rhizomatous roots and are increased when dormant. Cannas are not hardy. When frost blackens the foliage, chop it off, dig up the tubers, shake or rinse off excess soil, and store in a frost-free place. In early spring pot up divisions in loam-based compost. Give them light, air, water and warmth, and plant them out when frost is past. This gives them a flying start, so when their time comes they will burgeon. Ginger lilies are from the Himalayan foothills and don't need so much cosseting: braver souls leave them in the ground with soil heaped over their roots. It can look as though you're plagued by giant moles, but large enough mounds will prevent the rhizomes from freezing.

Dividing crown rhizomes

Big flowers of unbelievable beauty with layers of frilly petals or simple, single chalices – who can resist peonies?

Many peonies have a sweet scent and exquisitely formed flowers, the central boss of glorious golden anthers is often encircled by a ring of dark, dramatic splodges. Every garden needs plants like this for flower power. No matter how bold the shrubs, how naturalistic the planting, big blobs of colour are a huge draw in early summer. Peonies do it best. No other plant offers such a generous show, or performs so reliably.

From the mountains of Turkey and Greece, to the wooded valleys of China, Siberia and Japan, this race of plants offers gardeners in temperate zones a wealth of striking species. Generations of breeders have worked for hundreds, in some places thousands, of years to create cultivars galore that grow in a wide variety of soils and situations. Most cope even when a venue is not quite to their liking. A wealth of myth and legend surrounds peonies, nowhere more so than around their propagation.

Many of us are loath to lay a finger on our peonies. Much as we would like to split them, the received wisdom is that they don't take kindly even to being moved, let alone divided. It is true that they take a dislike to any upheaval, especially if their roots dry out. They may sulk for a season or possibly two, but if each piece contains a decent tuber or two and an 'eye' bud – an embryonic bud that will grow into a leaf and possibly a flower – they will have everything they need to flourish.

A mild spell during the winter offers the best opportunity to perform this operation, but if you miss the boat it can still be carried out in early spring, as long as the stems are short and sturdy and before any part of the plant becomes lanky. Here we dug up *Paeonia emodi* 'Late Windflower', and although the plant was lifted after the buds had started to shoot, each piece survived well and flowered the same season.

CUTTING A CROWN
Alongside inherited wisdom, myths and legends abound. Some can be off-putting, like the fallacy that if you divide peonies they will die. With a new bud – very obvious if you leave it as late as this! – and a good root, divisions thrive.

Dig up the plant during late winter or early spring. If it's too difficult to see what's what, wash it carefully so the buds can be easily detected. Slice through the crown with a sharp spade, a large knife or even robust secateurs, ensuring that each piece has several tubers and one or more eye buds. Take care not to sever the tubers from the crown, since they are the food source for the new plant. Text-book theory dictates that cut surfaces should be allowed to dry out by placing the divisions on a wire tray in a warm place and dusting them with fungicide. This is not an option if you garden organically, and frankly it's not necessary. If the tubers are healthy (you would not be propagating from them if they weren't), your tools are sharp and clean and the divisions sit back in fresh, well-drained soil or compost, there should be little chance of disease suddenly striking. Although divisions can be returned immediately to the soil, it is far better (especially with smaller pieces) to pot them up in a loam-based compost and allow them to establish well before letting them fend for themselves in the garden.

Wherever they are replanted, make sure they are not buried and that their planting hole does not become a grave. Whether you are placing new plants or replanting divisions, the practice is the same. Plant so that the crown of the plant and the new leaf buds are level with the surrounding soil. The main reason for peonies failing to flower is that they have been planted too deeply, and need to concentrate on staying alive rather than producing flowers. All peonies, especially forms of the familiar 'cottage-garden' peony (*P. officinalis*, and *P. fruticosa*), need

plenty of sun to ripen their tubers and produce flowers freely. Tree peonies need some shelter, but since they are grafted onto a rootstock they cannot be propagated in the same way.

Asparagus falls into the same category as peonies, with crown rhizome root systems, usually just called crowns when offered for sale. They can be divided with a sharp spade so that each piece has several buds and plenty of thick roots. It takes asparagus a few years to get into the rhythm of producing copious numbers of their scrumptious shoots, and this will be interrupted after such treatment, so the usual practice if you want more is to buy new crowns.

FIRST FOLIAGE
Although peonies are renowned for their voluptuous flowers, their foliage alone would be reason enough to grow them, especially when the first buds open in spring.

Bulbs, Tubers and Rhizomes

Dividing crocosmias

In mild areas of the countryside, you are almost bound to come across a somewhat incongruous orange flower.

As an outcast from gardens, *Crocosmia x crocosmiiflora* – or montbretia as it has come to be known – has made its way to cliffs, hedgerows and verges, especially along the south coasts of both Britain and Ireland. In our gardens, its colonizing habits have won the whole genus a bad reputation. The familiar (or perhaps over-familiar) montbretia was introduced as a novelty from the Lemoine nursery in France in the 1880s, and on its introduction was hailed as a wonderful new ornament for gardens. A few decades later it was thrown away in large quantities because of its invasive habit, not so much a garden-escapee as a garden outcast. Since then new hybrids have been introduced, some of them no more than a flash in the pan, some of them with real staying power. At times, scores of cultivars were listed by big nurseries, at other times they became unfashionable and sank out of favour. Recently they have had something of a

renaissance; old varieties have been resuscitated and new cultivars created.

Crocosmias fit well into the modern garden and style of planting. Their informal, wandering growth is ideal in naturalistic schemes: if you are the sort of person who likes everything in its allotted space, then perhaps crocosmias are not for you. They have a mind of their own and a wandering spirit. Their vividly coloured flowers (there is no such thing as a subtle crocosmia) lend a touch of the exotic, and their sword-like foliage in bright green – or occasionally bronze – makes a useful contribution from mid-spring onwards. In the low spring light, leaves are translucent and add an element of movement amongst static clumps of geraniums and other perennials. As the summer progresses the flowers supply an injection of hot colour that gets the garden pulse racing, and plants that wait until late in the season to do their thing are particularly welcome, especially if they are tough and enduring. If they are gracious too, and match the mood of the moment, they are especially treasured. But to give their all they need a little human intervention, and their response to sympathetic propagation is rapid.

All crocosmias grow from corms that produce new corms on their shoulders each year. A corm is a swollen plant stem and a storage organ that enables the plant to withstand extreme conditions – sometimes cold, sometimes heat. They look like bulbs on the outside,

146

but don't have a bulb's layers or scales on the inside: they're solid through and through. Gladioli, crocus and crocosmias all spring from corms. Mostly they are split and multiplied much as for bulbs (see pp.130–1). But crocosmias are a little different. They spread by making a fresh corm at the end of a chain of corms, resembling a string of giant beads (like those I used to wear in the late fifties, or early sixties), which build up in the soil to form matted clumps. In nature they journey around, constantly accessing new ground. In the confines of a garden, clumps tend to become congested, and consequently unproductive, left to their own devices, their flowers become poorer and more sparse. It's good practice to lift them every two years, breaking off the fat, new corms at the top and discarding the rest. Replant the young corms 5–10cm (2–4in) deep in ground refreshed with home-made compost or any other good organic material. This is best done in spring, although it can be done after flowering if all the top growth is left on to supply the corm with sustenance. If your soil is dry and thin, or if you are gardening in a cold part of the country, plant deep; sometimes the new shoots are clobbered by late frosts, but they always recover, and the fresh young vertical leaves are one of the delights of spring. Deep planting will result in bigger, better flowers and strong, healthy leaves that stand up for themselves proudly.

BREAK THE CHAIN
As soon as you pull away the top corms from a clump of crocosmia, you know that you have the very essence of the plant. As the leaves die down, the corms below swell.

Dividing tubers

Henry Pulling, the retired civil servant in Graham
Greene's *Travels With My Aunt*, grows dahlias.

In a monochrome life bereft of colour
and spontaneity, they are his
indulgence, his secret excitement,
his only connection with the thrilling
– until his aunt comes along. Middle-
class Henry was an exception to the
rule: once upon a time, dahlias were
the reserve of the allotment holder.
They were a weekend, after-shift sort
of a plant for the working man to
perfect and display on the show bench
– or just to indulge in for their vivacity,
a glorious distraction from day-to-day
reality. Many gardeners have frowned
upon them, but times are changing.

MAKING THE MOST OF SPECIAL SEED POTATOES

On this side of the Atlantic, we plant out whole seed potatoes for our new crop, but in the United States the practice is to cut them up and plant the pieces. As long as each has a short sprout (two eyes are usually recommended) and a chunk of tuber about 4–5cm (1¾–2in) each way, this should work perfectly. It is sometimes recommended that the sections are allowed to dry for several days before planting, as a precaution against infection, but they can be planted immediately if everything is clean.

GET A HEADSTART
If you are impatient for big dahlia plants, rather than raise more from basal cuttings, you can cut through the collection of tubers and replant each chunk individually before encouraging them into growth.

Up and down the country dahlias are more and more in evidence in our gardens, and more and more of us are starting them from tubers or buying fully grown plants in bloom at flower shows or garden centres.

They're such joyous plants, it's difficult not to indulge, but dahlias will never quite gain the respectability granted to many other exotic flowers. There is something much too forthright, too upfront about these glorious blooms for them ever to be considered quite 'proper'. Yet their very blatancy has put them at the top of many a must-have plant list.

Dahlias are best divided in late winter and early spring. Uncover tubers that have been asleep, packed in bark, and divide them with a sharp knife. Plants older than three or four years should yield several pieces, each with part of the crown and several tubers attached. Dangle them into clay pots and pack around with peat-free compost. Water well once and put them in a warm, sunny place. Once they sprout, extra basal cuttings can be taken.

NATURAL GENIUS

Bulbs are extraordinary: they move in the soil, clone themselves and walk.

ERYTHRONIUMS

The roots of a dog's tooth violet pulls the plant to the right depth in the soil. This is a characteristic seen to some extent in many bulbs and corms, but erythroniums take the trait to such extremes that they are slippery customers to grow in pots. This is because they will slither out through the holes in the bottom of the pot when you're not looking. The important part is to plant the bulbs the right way, with their shoulders and the contractile roots pointing up.

SNOWDROPS

Christopher Lloyd always maintained that one of the earliest-flowering, most outstanding, and fastest colonizing snowdrops, *Galanthus* 'Atkinsii', did not set seed, but increased solely by means of its bulbs dividing spontaneously. This stately, tall cultivar is a hybrid, but it spreads so rapidly that I was convinced it must seed itself as well as increasing by its bulbs. How could any plant establish such large colonies from just a few bulbs, and in a matter of no more than couple of years, without spreading by seed? But of course Christo was right, and the success of *Galanthus* 'Atkinsii' is a signal example of just how fast some bulbs can spread even without having the advantage of setting seed.

ALLIUMS

Many of the alliums produce bulbils in their flowerheads. Most notable is the tree onion, *Allium cepa* Proliferum Group, whose bulbils are sold as silverskin or cocktail onions. It is also called the walking onion, because the stem dries and falls, depositing the bulbils on the ground where they grow and produce a flowerhead full of bulbils, which in turn falls when the stem dries and deposits the bulbils on the ground, and so on, moving across the ground in neatly measured strides.

LAYERING

On the way to our kitchen door here at Glebe Cottage is a beautiful shrub. In the winter its shape is impressive, arching and elegant, but it's still easy to overlook in its bare state. In late spring it's another story – nobody, not even non-gardeners, can ignore it. Everyone stops to take in the sight of the pure white chalices that crowd its boughs. It is *Exochorda macrantha* 'The Bride' and it was one of the first shrubs we planted here when we came. Having gazed at pictures of it for years, it was thrilling to be able to plant one in our garden. It has had a chequered career over the years, moving to several sites in the garden and once taking a trip to London in the back of my old Mercedes van to take pride of place in one of our displays at a Royal Horticultural Society Spring Show.

For the last 15 years or so we have left it alone, and it has rewarded us with a magnificent show, but two or three years ago something worrying started to happen. The branches on the west side of the shrub were nowhere near as well-clothed in flowers as those on the east, nor were the leaves that followed anything like as lush. The problem got worse, yet on the other side of the shrub the growth and general health seemed as good as ever.

On close examination it became clear that while one side was on its way out with peeling bark and dwindling twigs, the healthy side had surreptitiously layered itself and was now operating from an entirely new root system. The old, dying branches have now been cut out to give a fighting chance to what is in effect a new shrub.

So my exochorda has layered itself and has had a new lease of life as a result. You will often find examples of this happening spontaneously: as far as the plant is concerned, this is an insurance policy, another method of prolonging its life. In historic gardens and arboreta you can see old specimens of Western red cedar (*Thuja plicata*) whose branches have sagged, rooted, and formed a grove that can survive the death of the central parent. No doubt the first attempts at layering shrubs were made as a result of gardeners observing trees and shrubs naturally doing it for themselves. Long ago gardeners investigating branches that having arched down to the ground seemed to have made roots must have tried to emulate what they saw.

The process is extremely simple, and all that is needed to successfully layer a shrub or tree is patience. A whole year may well elapse between the moment when you push the stem

FRUIT FOR FREE
Push pots of good compost into the soil and encourage strawberries to root into them. When they are well rooted, their umbilical stems should be severed.

into the ground and the day it has finally established enough roots to be severed from its parent and taken off to a new life on its own. But nothing in gardening is instant, and there is no finality, which is surely partly why we all become so involved: we are part of a continuing process.

Probably the most spectacular tree in my garden is a voluminous *Cornus* 'Norman Hadden', named after the great gardener of Porlock, Somerset, who hybridised this tree from seedlings

he had grown. My own tree came from Rosemoor, the RHS garden at Torrington not far from Glebe Cottage, where I had my first meeting with the tree three decades ago. The garden was then still owned by Lady Anne Palmer (now Berry), and it was always a source of inspiration for all who visited. On one late spring trip I came face to face with the tree, its branches decked in huge white bracts, and I knew I just had to have one – any keen gardener will recognize the syndrome.

And I was in luck: there was just one plant of it in the plant sales nursery, and it had to come home to Glebe Cottage. It was by no means a perfect specimen, sort of sprawling over the edge of the pot, and more horizontal than vertical. Lady Anne made no apologies: it was a layer, it had made itself, it was the only one in the place, and I could like it or lump it. Hobson's Choice. I liked it, and paid my £4, which was a huge amount of money then. I planted it carefully as soon as we got home (it wasn't even established in its pot) and went out to talk to it every day to persuade it that the garden here was where it was destined to live happily.

This lovely tree has increased slowly over the decades and now graces the garden right through the year, occupying a central position, much admired by all who see it. And yet, can you believe it, we have never tried to layer it! Well, that can be remedied as soon as this chapter is finished.

NEW FROM OLD
The original plant of this lovely exochorda is on the way out. It's for the chop, but luckily it has layered itself and the new plant is full of vigour.

Making layers

Occasionally when weeding around a shrub or when planting bulbs in a quiet corner, you come across a suprise bonus.

A low-lying or arching branch from a shrub has taken root quite spontaneously and lo and behold you have a new shrub.

As in so many country gardens, there is some honey-fungus at Glebe Cottage: it's not too much of a problem, but every so often some poor shrub or tree is stricken. One of my first acquisitions, a few weeks after moving to Devon, was a *Mahonia japonica*, purchased from my friend John Turner who had a small nursery close to Glebe Cottage and took his wares to the Pannier Market in South Molton. I was teaching then and living in a rented farmhouse with no garden, but the big, glossy leaves and racemes of glowing yellow flowers, scented like lily of the valley, were too irresistible. After school the mahonia was carried triumphantly home in my bicycle basket and placed in the glass porch. It was the middle of winter, and soon

afterwards the southwest was under snow, the worst blizzards experienced for 30 years. Later that year we, and the mahonia, moved to Glebe Cottage. We all loved our new home, and the mahonia went from strength to strength. Ten years later it started to lose leaves and vigour, and eventually died from honey-fungus, but one branch had layered itself. Twenty years on, this branch is a fine, handsome shrub, much visited in the winter for big sniffs of its uplifting perfume.

The best time to layer shrubs is during late winter and early spring. Any branch from almost any shrub can be layered, but try to find branches that 'suggest' themselves. You are looking for young, strong, whippy growth that can be brought down to ground level without any danger of snapping. The branch must be long enough for you to be able to bury part of its length and still have a good

EASY DOES IT
Young, supple stems often 'suggest' themselves as candidates for layering. To give the best chance of rooting, prepare the ground well, wound the stem, make sure it is stable, and give it support to grow straight. Then wait.

length beyond the buried stem to
make the first trunk of the new shrub.

The idea is to stop the branch
growing upwards – producing apical
growth with more stem, leaves and
new shoots – and to persuade it to
grow downwards and make roots
instead. Pull down the stem to
make sure it is suitable and to get a
rough idea of where it will be buried.
Prepare the soil around the area, if
necessary by turning it over gently
and incorporating drainage material
– compost and grit – if it needs
improvement. The idea is to create a
soil environment that will help rooting
to take place. Dig a shallow trench and
add extra grit to its base.

Trim any leaves or side shoots from
the part of the stem that will be below
ground level. You can lay the stem in
the trench just as it is, but wounding
it will help induce it to make roots. It
sounds mean, but it won't do it any
harm. Sometimes it is suggested that a
ring of bark is trimmed off with a sharp
knife in the area you want the stem to
root, or that a piece of wire is twisted
around the stem so that it cuts into the
bark at the same point. I prefer to cut
into the stem at an acute angle, sliding

in a matchstick, or any other piece of clean wood, to keep the cut open. Once the cut is made the stem is more liable to snap, so it needs to be lowered into the trench gently. Once there, secure it with a staple made from malleable wire (florist's stub wire is ideal).

Backfill the trench, firming it down well. Some gardeners keep everything in place with a stone; the only disadvantage of this is that it can prevent water from permeating the soil, and the layer needs moisture if it is to make root. The extension growth (the bit sticking out on the other side of the trench) will be horizontal at this stage, so push in a stout cane or stick beside it and tie it in with twine to bring it upright. Water the layer well, then forget about it, apart from checking occasionally to ensure the ground around it is moist.

It won't be until the next autumn that an adequate root system will have built up. It is very exciting to investigate tentatively and discover white roots just below the surface of the soil, but if you think the root system is insufficent to fend for itself, give it a bit more time, perhaps until the following spring. Loosen the surrounding earth so that you can see what is going on and with a sharp pair of secateurs sever the stem on the shrub side and lift. You can replant it into the ground or a pot, but in either case make sure in advance that everything is ready for your new plant.

VARIATIONS ON LAYERING

There are a few other ways to make layers, but they all work on the same basic principle of wounding a stem and provoking it into making roots before severing it from the parent plant.

Stooling or French layering is useful if your plant is vigorous, but has no shoots that can be pulled down to the ground. You simply steel yourself and cut all stems back to the base in spring. The resulting new stems will be flexible enough to be pulled down the following winter, and you can pin them out like the spokes of a wheel. Heap soil over them as new vertical shoots grow, and a year later you will have new plants.

Air layering is another way to deal with the lack of low shoots. Instead of bringing the shoot to the rooting medium, you bring the rooting medium to the shoot. Wound a shoot where the wood is a year or two old, pack damp sphagnum moss round it, wrap it with black plastic – a strip torn from a bin liner is ideal – and seal with tape above and below the wound. Leave it for about a year, checking now and then for roots, and when there are enough remove your new young plant.

GREEN SHOOTS
OPPOSITE It's always so exciting to dig up a well rooted layer. **ABOVE** Blackberries, logan berries and their ilk can all be tip layered, replicating exactly what the plants do naturally by burying their tips. A new shoot will be formed at this point, and then the main stem can be severed.

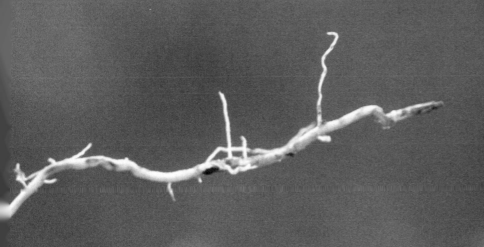

DIVISION

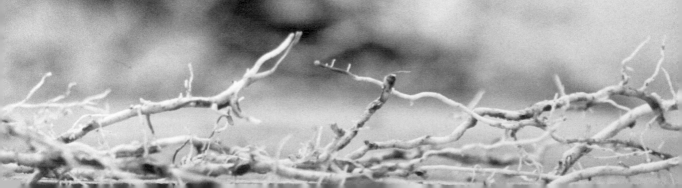

There is something slightly unsettling, yet at the same time deeply satisfying, about dividing plants. It seems rather heartless to dig up a healthy clump of hemerocallis or lift fat rosettes of primulas that have just graced us with a fine show of flowers; then to tear apart the clumps into smaller and smaller pieces simply adds insult to injury. But to carefully replant them in a specially chosen place with lashings of compost, to settle them in with a can of water through a fine rose and to tuck them up in a blanket of mulch – this is a labour of love.

For many of us, division is our first experience of propagation, although we may not recognize it as such at the time. On acquiring our first garden, those gifts that arrive in carrier bags, newspaper parcels and cardboard boxes from gardening neighbours or friends and family are for the main part divisions from their own plots. And in many cases the divisions that are being passed on to us will come from plants that arrived with our benefactors as divisions too!

These are the most immediate examples of plants being passed on through time – literally and physically. We may send seed or swap it, and it can travel thousands of miles from the place where it originated to its new home, but there is always choice and the chance involved here: the choice about whether or not to sow it and the chance of whether or not it will germinate. In contrast, giving and receiving divisions of plants is a much more immediate, over-the-garden-wall sort of business.

Dividing herbaceous plants is one of the oldest ways of making more, and it has much to recommend it. It is simple and straightforward, and it has few of the risks, problems and dramas that attend some other methods of propagation. Providing your divisions are made at the right time and are replanted sensibly, nothing much can go wrong with them. It is the most palpable of all methods, and the most physical; there is no option other than to get stuck in, and gardeners can immediately see and use the fruits of their labours. New plants are made instantly and should grow and flower in the same year from spring division, or in the next season from division in the autumn.

Back in the heyday of the grand herbaceous border, the entire contents of a border would be lifted and divided every second or third year. The old, woody centres of clumps would be discarded, and the whole empty bed

KINDEST CUT
LEFT Hostas can become so densely packed they need to be sliced apart with a spade.
RIGHT Fast-growing asters can quickly become congested; cut them up into smaller chunks and discard the centre.

dug over and have muck incorporated into it before the divisions were replanted. If the borders were exceptionally long, they would be divided into sections and each section tackled on a rota, a bit like painting the Forth Bridge. Although this mammoth work and upheaval would have given the gardeners more plant material to use elsewhere, the main reason for the process was to give new vigour to the border.

Today, this procedure might be regarded as mechanistic. There are some herbaceous plants that increase in stature each year and therefore need time to reach their glorious maturity, while others need to be split and replanted regularly to guarantee their vigour. Although we no longer divide our plants as a matter of course, it is still good practice to split herbaceous perennials as soon as they start to show any signs of going downhill.

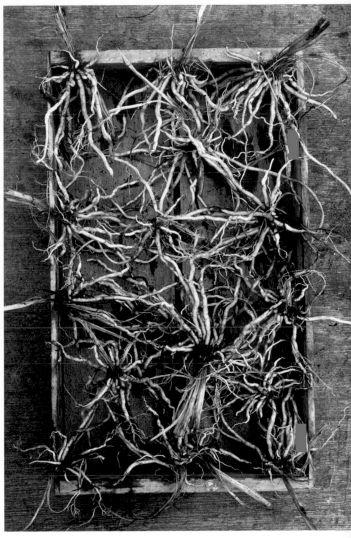

In nature, plants constantly seek out fresh ground, moving away from the old, woody centre, and we exploit this propensity when dividing plants. If a clump has had enough room while it has been growing, it will always move out from the centre, and if a clump gets a bit thin in the middle, it's time to act. Often your plants will let you know in other ways when they are in need of splitting: when the stems of clumps of phlox, for example, start to look spindly and become congested, and flowers become smaller, it is best to dig them up and divide them into smaller pieces, each with no more than three or four stems.

As well as providing masses more material, the process of division rejuvenates the plant, promoting new, vigorous growth. It promotes your own vigour too: even in the depths of winter, outer layers of clothing will be abandoned as you get to work.

PULLED APART
**LEFT A single clump of grass like this deschampsia can yield scores of smaller plants.
RIGHT A few plants, like this ranunculus, almost fall apart.**

Deciding to divide

The best time to split clumps of any plant is during spells of mild weather, when the plant is dormant.

Most perennials can be divided in spring or autumn. Which season you choose depends partly on your own convenience, partly on the area where you live and the prevailing climatic conditions, and partly on the plant – and the plant's needs come first. A few plants, such as asters, resent autumn division and often deteriorate and die when subjected to it. Unfortunately, there are no hard and fast rules, but as a guide it is best to divide plants that flower early in the year in the autumn and to avoid dividing late-flowering plants until the spring.

In autumn, the soil is still warm and roots can start to take hold; many plants will have been cut back earlier to make way for the autumn show. Since they have been deprived of the possibility of making seed, their energy will be concentrated on making roots. Late flowerers, on the other hand, make the most of their root growth and form new shoots in the spring. After flowering they need a rest, and may object to being asked to re-establish immediately after they have given their all. For plants in cold, wet, heavy soils, it's also usually preferable to wait until spring. The optimum time is when the soil has begun to warm up, yet the plant is only just beginning to stir: early spring in the south and mid- or even late spring in cold regions.

If possible, the sites where new divisions are to be planted should be prepared before the plants are dug up,

so that there is no delay between the roots leaving the soil and getting back down there again. Roots should never be allowed to dry out, so if it starts to rain or bad light stops play, ensure that they are packed around with damp soil in boxes or heeled in to a trench in the ground.

Because clumps spread outwards, the youngest, most promising material for new plants will usually be on the outside of the clump. Some roots can be split easily by simply delving in with your hands. Recalcitrant clumps may need to be tackled with a knife or, in some cases, a saw. I have been known to split old clumps of miscanthus with an axe, but such drastic practices only need to be used in extreme cases – usually when a tough plant has been ignored for far too long.

Leaves can be trimmed back to reduce transpiration and let the plant concentrate on getting established. The majority of herbaceous perennials have fibrous roots that can be trimmed back, but unless they have been damaged, it is unwise to trim the roots of fleshy-rooted subjects, particularly kniphofia and many plants in the lily family, as well as members of the buttercup family such as *Ranunculus aconitifolius*. Try to maintain the whole root system to prevent rotting, despite the fact that this will mean you need to dig a deep enough hole to accommodate it, and make sure your division has plenty of rich organic matter to wallow in.

DOUBLING UP
Some plants need pulling apart, others need forks and a few need a blade, but the principles are the same for all of them.

Straightforward division

Many perennials make big clumps with dense crowns and copious thick roots – it's hard to go wrong here.

Splitting these clumps into smaller pieces and prising the old woody middles away can be an energetic procedure, but since it can be undertaken on a chilly morning in early spring it's just what's needed to get the circulation going. Division is probably the most physically demanding method of propagation. It's also hugely satisfying to lift an enormous clump and see it, through your own efforts, broken down into smaller and smaller chunks. Psychologically, too, it is gratifying: direct action, apparently destructive, which is in fact positive, constructive and helping to create new life.

Division is the usual form of propagation for day lilies. Although a few species set and ripen seed, and new plants can be successfully raised from this seed, the great majority are sterile and can only be increased by vegetative means. After flowering is an excellent time to divide them, but it leaves large, messy spaces wherever they have been lifted, and even if foliage is cut back when replanting there is an unhappy hiatus at a time when everything should be burgeoning and at its peak. Dividing them in the spring is a much better idea: they are early into leaf so you can see what you are doing.

Dig up the whole plant with as much soil attached as your back can lift. Put the whole clump onto a flattish surface, ideally unoccupied soil. If you're working on a path or grass (we have no lawn at Glebe Cottage), old compost

bags or sacks laid around you will prevent soil and roots making a mess.

With big clumps – and most are – plunge a border fork vertically into the centre of the clump and another one of similar size in behind it so that the two are back to back. Push the handles away from each other at the same time and gently lever the clump apart. Break down the clump into chunks, with two or three growing points firmly attached to their roots. Shake soil off gently to see what you are doing during the process.

Several herbaceous members of the rose family, including astilbes, filipendulas, aruncus and geums, make woody crowns that can be split using the two forks method. Each clump will have several resting buds. The procedure is the same as that for hemerocallis, but in most cases not as bicep-building. Beneath their woody crown, most of these rose relatives have fibrous roots, and in their case it is worth trimming them neatly.

Old woody centres should be discarded; it is easy to identify the old, spent parts of plants. Not only will any buds produced there be shrivelled, but the roots that try to support them will be thin and congested (almost woolly) when the soil is shaken from them. Chunks of the fresh, young material from the edge can be broken up into goodly pieces, each with three or four buds. Put them into a plastic bag until you can deal with them, so that the roots remain as plump as possible. Though it is a counsel of perfection,

SPRING WORKOUT
To maintain flowering performance, some plants need dividing every few years. With plants on the scale of hemerocallis, dividing them can also double up as a keep-fit class.

whenever you divide a plant try to ensure that its future home has been prepared first, whether it be pot, garden or nursery bed. The less stress, the better it will settle.

WASHING ROOTS

Some hemerocallis, especially the modern, large-flowered cultivars, can easily be pulled apart; rinse the roots as you are going along to see where divisions can easily be made. Washing roots is always worthwhile when it's difficult to see new shoots or resting buds. Swishing roots around in a bowl or bucket of water reveals all. You can use warm water, but never hot – you don't want to cook the roots, but a happy compromise can be reached to

make conditions more comfortable for you without damaging the plant. For the most part, however, you will not need to resort to washing roots.

Many other plants lend themselves to this kind of division; other members of the lily family, such as kniphofia and paradisea, have separate growing points joined into one crown. Neither makes such robust crowns as day lilies, and there is more danger of damaging them if you are too cavalier. Lift them with as much root as possible, shake off soil or wash the roots and push and pull the growing points apart. Only trim roots that have been damaged, then replant in generous holes plenty deep enough to accommodate the roots without bending or folding them.

171

Dividing fibrous-rooted plants

The wild primrose (*Primula vulgaris*) has given rise to hundreds, possibly thousands, of cultivars and sports.

Primulas have been in cultivation for many hundreds of years. They are among the most popular of garden plants not just at home, but also in North America and New Zealand. Gerard's *Herbal* mentions several double forms already considered valuable garden flowers in the 16th century: 'Lilacina Plena' or 'Quaker's Bonnet', which does well at Glebe Cottage and in thousands of gardens up and down the country, is one of the prettiest doubles. It has a lax habit and a profusion of very refined lavender flowers with rather weak stems that give it a 'delicate' air; it is far more elegant than some of the more modern doubles that are now increased in vast quantities by micro-propagation.

Though primulas are child's play to grow from seed (see pp.32–5), the only way to maintain a healthy number of special varieties is by division, and it needs to be done frequently. Most of the doubles are sterile, and for the amateur gardener division is the only way to make more. Varieties like *Primula* 'Dawn Ansell', with a posy of pure white petals contained within a 'jack-in-the-green' ruff, are very desirable, but most effective in groups. Several plants can be used to weave their way between dark hellebores or contribute to a green-and-white planting with *Leucojum aestivum* or *Pulmonaria* 'Sissinghurst White'.

Gertrude Jekyll grew great drifts of polyanthus in her coppiced hazel woods and sold the excess when they were split each year. She described the best time to divide them as 'when the bloom wanes and is nearly overtopped by the leaves'. Miss Jekyll depicts the scene in the middle of her own wood at this time with one garden lad lifting the primulas and bringing them to her where she was ensconced on a chair trimming roots and pulling off excess leaves, before another boy took off the doctored divisions to replant.

The best time to carry out this operation is on a cool, damp day when the flowers are fading. Early to midsummer is a good time, but not when the sun is shining bright. Strip off some of the older leaves too, so that the plant is not overloaded trying to keep big, old leaves alive at the same time as trying to make new roots.

Each of the rosettes should have its own root system, but you will need to trim off any old, woody roots below the new ones. (If you are greedy, even these old pieces will regenerate if they are plunged into a deep seed tray or a pot of well-drained compost.) With a sharp knife, trim the roots to about 10cm (4in), the length of an average palm, so that they are all at full stretch when replanting. Remove any old leaves cleanly and replant, taking care not to bend or fold back the roots. They thrive in cool, humus-rich soil and appreciate a thorough mulch with leaf mould or home-made compost.

If there is too much going on where your divisions are to live permanently,

DIVIDE AND RULE
In a garden setting, primulas can have their style cramped by other perennials. To keep them growing heartily, lift the whole clump just as flowering finishes and pull it apart with your hands, shaking off soil and discarding old woody crowns. Having trimmed the roots, line out the pieces.

plant them out temporarily in a little nursery row or in the veg garden to establish. We grew our best Barnhaven primulas and polyanthus underneath blackcurrant bushes, where they appreciated the rich diet of old muck and the peace and quiet. Although most primulas are tolerant of a wide range of sites, they are in their element amongst shrubs or under the canopy of deciduous trees.

OTHER SUITABLE CANDIDATES
The same methods should be applied to all fibrous-rooted perennials. Some, such as hardy geraniums, may not have such distinctive rosettes as primulas, but the principle still remains the same. Trimming the roots encourages fine, fibrous, feeding roots to form. Many cranesbills have fibrous roots: *Geranium oxonianum*, the most commonly grown pink geranium, is typical and once it is lifted can easily be pulled apart. Some of the larger cranesbills, such as *G. psilostemon*, need to be carefully separated at their crowns and teased apart, occasionally with the help of a knife. Being a species, *G. psilostemon* sets seed too, but dividing it means you instantly have sizeable chunks to start new planting schemes. For most species, seed is the easiest option, while a few such as *G. maculatum* make woody tubers (see pp.148–9), and for many of the new cultivars basal cuttings are the only way to make more. But for the great majority of these popular plants, division is the way forward.

Division is also the usual means of making more epimediums. The majority of these desirable plants have woody crowns with a mass of

fibrous roots. Lift the whole clump, and shake off any excess soil, then pull them apart. It may be necessary to use a sharp knife to start off the process, but try not to sever the dormant buds, because these are what will make the new season's growth. Each piece needs four to six buds to get going and make a viable plant in the next season. Mark their positions well if they are planted back into the garden, especially if they are deciduous types, such as the mouth-watering cultivars of *E. grandiflorum*. Leave on the leaves of evergreen epimediums, such as *E. versicolor*. Chunks may need to be anchored by a strategically placed stone or a giant 'staple' – florist's stub wire is ideal for making these.

If you are planting your divisions among tree roots, which are a natural place for these plants to grow, excavate carefully to find pockets of soil and then add plenty of humus-rich material around the epimedium's roots. It is always worth getting the plants off to a good start; in nature, both tree and epimedium would have grown up together, so in a garden situation where the tree is already established, it seems unfair not to give the epimedium a bit of extra help.

STOLONIFEROUS PRIMROSES

Not every primula grows like a primrose. *Primula sieboldii* makes a dramatic spring display of broad, bright leafy clumps, bespangled with flowers whose fragile beauty belies the plant's toughness. It grows quite differently from *P. vulgaris*, making small stolons or pips, each of which makes a separate flowering shoot. In the wild this species grows in damp meadows, and in the garden, just as in nature, it will put up with competition from other plants that overtake it after its spring extravaganza has ended.

This is an obliging plant, that withstands low winter temperatures and asks for little, apart from moist soil and an occasional mulch, which should guarantee its gentle spread. To divide it, whether from pot-grown plants or in the ground, wait until the pips have started to make fresh root at the beginning of the winter, then move in and pot them up, just covering them, several to a small pot in good, leafy compost. Let them establish and when they are showing first leaves and a notion of flower bud, plant them out – or put them in a decorative outer pot to admire them, as the Japanese do, close up as they come into flower.

TRY THESE
The methods for dividing geraniums like *G. oxonianum* and *G. psilostemon* (above left and centre), primulas like *P. sieboldii* (above right), and epimediums like *E. grandiflorum nanum 'Freya'* (opposite) differ slightly, but all are easily divided.

174

Plants with thong-like roots

Pulmonarias, omphalodes, symphytum and the borage family in general fall into this category.

RESCUE MISSION
LEFT Clumps of pulmonarias can become congested or just tired out under the roots of shrubs or trees. Lift them, pull them apart, trim and replant them.

MOVING ON
OPPOSITE Now they've been saved from the increasing competition of the *Cornus controversa* 'Variegata' roots, these pulmonaria divisions are off to a new home.

Pulmonarias flower from as early as Christmas (with us!) right through to mid-spring. The midst of their quiet time, before they have made new buds, is the perfect opportunity to make more of them. Pulmonarias will sometimes seed themselves around, but there is no knowing for certain how these seedlings will turn out.

Pulmonaria longifolia and its cultivars can be increased by root cuttings ssbe guaranteed because each cutting will be a clone of its parent, but this doesn't work for all of them. The special forms of *P. saccharata*, *P. officinalis* and *P. angustifolia* can only be increased by division. At Glebe Cottage we use pulmonarias

extensively in the shadier reaches of
the garden, but we also have some on
the sunny side of the garden. Here they
are planted amongst other perennials,
amongst whom they happily nestle for
the rest of the growing season after
they have finished flowering.

This is a good place for collections
of shade-loving spring bulbs, too. In
their famous garden at Glen Chantry
in Essex, Wol and Sue Staines use the
flowing borders filled with a succession
of amazing herbaceous plants and
grasses to house their fabled collection
of snowdrops. They make an ideal
partnership: I've copied this practice
in one of our borders and in one patch
the snowdrops are planted with the
pulmonaria 'Dorah Bielefeld', astrantia
and sanguisorba.

By autumn, pulmonarias will have
moved in several different directions
making new roots. They are lifted
gently with a small border fork and
lowered into a box. Carefully separate
the pieces and break off the old roots.
Trim the leaves right back: this seems
brutal treatment, but it reduces each
piece to its lean and vital self. It has
everything it needs not just to survive,
but to grow into a fine new plant,
providing the gardener gives it what
it needs: a shady site with good soil,
a generous helping of organic matter,
home-grown compost, leaf-mould or
the like and a good drink. Plant the
pieces a handspan apart. The divisions
can concentrate on taking root first
and will produce fresh new leaves
later. They may not produce flowers
in their first winter, but they will build
into a fine, ground-covering swathe
of rich green, silver-splodged foliage,
which is almost enough in itself.

DIVISION BLUES
**One clump can make
scores of pieces,
trimmed to their
bare essentials
(opposite). These are
carefully replanted
with extra compost
to get them off to
a flying start (left).
For cultivars like *p.
angustifolia 'Blue
Ensign'* (above) this
is the only way to
make more.**

179

Dividing solid crowns

Plants with dense crowns, such as hostas, are very difficult to divide by levering them apart.

They are solid and thick with roots and a plethora of new buds that will be next year's leaves. Plunging two forks into them and attempting to lever them apart may result in damaging the plant, knocking off new buds willy-nilly and defeating the object of the exercise. The best procedure is to leave the whole clump in situ and, in early spring when the new buds are visible, to cut out chunks using a large knife or a sharp spade. This can be done in the autumn too, just as the foliage finally begins to fade. Hostas often put on a fine show of autumnal colour, creating glowing, golden patches in the mellowing scene. The chlorophyll that made their leaves so green earlier in the year is broken down and the goodness from the leaves is drawn down into the crown of the plant.

If you want to maintain the volume of the main clump, rather than making several new, evenly-sized pieces, treat it like a cake and cut out one or two slices before returning the mother clump to the soil. Substitute a sharp spade for your cake knife, aiming it between the middle of the clump and the outside edge. Keep the blade of the spade absolutely vertical. You can remove about one-third of the plant, cutting in a few hearty chunks. The original will soon get over the assault, especially if all the spaces created by removing the other third are filled in

with good compost. The divisions can be replanted in their new home, which may be in the ground or in containers. Gertrude Jekyll often grew her hostas (although in those days she would have known them as funkias) in pots, where the architectural qualities of their leaves could be fully appreciated.

It may seem a brutal method, but really it is very civilized butchery: both the divisions and the mother plant seem pretty oblivious and usually respond by making energetic new growth in the spring. One factor to bear in mind when dividing hostas, though, is that pieces chopped out of a parent plant will be juvenile in their first two years and only then develop to adult proportions with leaves on the same scale as the mother plant.

PIECE OF CAKE
Slice of hosta, anyone? Taking out large chunks complete with strong roots and plenty of resting buds is probably the best way to split hostas.

Exceptions to the rules

Some plants defy categorisation when it comes to division, and a few almost fall apart in your hands.

Solomon's Seal (*Polygonatum* x *hybridum*) is a hard-to-define sort of a plant when it comes to propagation. It is a British woodland native that is closely related to lily of the valley, but with a completely different rootstock. It makes large, solid rhizomes, each ending in a fat bud that becomes the flowering shoot. The stems, up to 60cm (2ft) tall, are swathed in elegant leaves and arch gracefully towards their tops where a series of pale bells, touched in green and delicately perfumed, are suspended. There are variegated versions of the plant too, one with striped leaves and another whose leaf margins are cream.

Solomon's seal can be split or divided during its dormancy or even in autumn, with its leafy stems still intact; they will gradually die back, transferring all their goodness to the tuber. Each division should have a chunk of rhizome with two or three embryonic buds. There are other, less well-known members of the family too. *Polygonatum verticilliatum* has tall, straight stems with tiny bells in whorls around the stem. It spreads quickly and makes an elegant vertical storey above low-growing woodlanders, such as wood anemones and bluebells.

PLANTS THAT FALL TO PIECES

A few plants seem to divide almost spontaneously. The crowns of autumn-flowering gentians and some rockery campanulas, for example, are composed of separate buds or shoots, each with an independent root system. Conveniently, one of the most beautiful of spring flowers, *Ranunculus aconitifolius* and its double form 'Flore Pleno', also behave in this way.

PULLING POWER
Polygonatum hybridum (far left), lobelias like 'Russian Princess' (left) and *Ranunculus aconitifolius* (above and right) can all be pulled apart. Plant the pieces out direct or pot them on until they are established.

Many other plants 'spontaneously' divide. Why bother to propagate them then if they can do it themselves? Within a couple of years of being planted, a helenium will have increased exponentially so that what started as a single crown may now be composed of ten or more pieces. Providing it is well fed, it will continue to increase, but the clump will become more and more dense, and eventually the crowns of which it is composed will begin to dwindle and wain. Being able to exploit what the plant does gives gardeners a source of valuable new plant material; but it is not entirely selfish and benefits the plant too.

In spring, when each new crown has made fresh white roots, the whole clump can be lifted easily and the stumps of last year's flower stems can be used to prise apart each individual crown. 'Prise' is perhaps overstating it a bit: the pieces seem to want to fall apart and, should you care to start off new clumps of a respectable size, the best policy is to replant three or four individual pieces close together to form a reconstituted clump. If the new planting place is well worked with added compost and a sprinkle of organic food, and you leave a handspan or so between each piece, they will thrive and make a unified clump in the same year.

Heleniums make big, bushy plants at summer's peak and, if they are dead-headed, will continue to expand right through the autumn. From one big old plant you should get enough material for a short swathe and bearing in mind that they tend to have bare legs and knobbly knees, if they are planted with a shorter, better-clothed companion (rudbeckia or one of the handsome

evergreen euphorbias such as *E. martinii* spring to mind), the swathe or wave can become more voluminous.

Perennial lobelias are exciting plants. Some have suave, dark, deep crimson-red foliage, verging on bronze, and make a series of rosettes that are lavish in their own right, but are really a foil for tall stems of flowers in some of the most outrageous colours in the garden. 'Pink Elephant' is shocking pink, fifties lipstick, 'Tania' is rich, royal purple and 'Russian Princess' combines dark foliage and scintillating magenta flowers. Sometimes, if there are only one or two, they can look a bit incongruous, but go to town with them, plant them *en masse* and they make a splendid show. Their wild forerunners live in damp soil and make superficial roots feeding on the rich moist soil of bogs or beside streams.

After a summer outside, clumps will have expanded into a series of rosettes that can easily be divided and potted. Once established, they can be transplanted into beds and borders. You might even get those slender spires of blooms this season that are as good as anything you could buy.

IRISHMAN'S CUTTINGS

In theory, asters, or Michaelmas daisies, are the simplest of plants to increase. The traditional method is taking so-called Irishman's cuttings, which involves pulling away sideshoots that have already rooted. These can be potted up individually, but are usually transplanted directly into the soil. Since the operation is usually performed in the spring, they settle down well in the warm soil. Such plants tend to make loose crowns, so pieces are easily detached. Where the crowns have become congested, it is often necessary to lift the whole clump, shake off or wash off the soil, and pull it apart. In some of the species and cultivars that are close to the wild plant, root growth may become solid and woody. An old bread knife or pruning saw should get to the heart of the matter, and crude surgery can be used to remove any really old wood. Remnants of old flower stems can be used as levers. From one congested clump, many more can be made. To be viable, each piece needs at least a short stem and a piece of rootstock with some new roots coming from it. Plant several a few inches apart and pointing in different directions to make a good, random clump, hopefully full of flowers in the first year. When these new plants mature, the process continues.

185

Dividing grasses

Dividing grasses, from dainty deschampsias to monster miscanthus, can seem tricky: when, how, what size?

Grasses are a group on their own, and their propagation has some very exacting requirements, although essentially it is straightforward.

First and foremost, they should always be divided in spring, when their roots are beginning to grow again after retreating into the central clump with the onset of winter. Dividing grasses in the autumn and winter is asking for trouble; we cannot expect a plant intent on dormancy to summon up the energy to make new growth/roots, especially when it is cold and wet.

Many of the softer grasses, such as *Deschampsia flexuosa*, can be lifted from the garden or removed from their pots and divided. Gently shake off as much compost or soil as possible. Pull apart by plunging your fingers into the centre of the plant and making several pieces. Pull these apart into smaller pieces, removing any debris, until each piece consists of several basal shoots and a quantity of fine root. Any old roots should be discarded, and it's easy to distinguish between new and old: new roots will be fresh and pale, often white, whereas old roots are skinny and dark. Pot into modules and place on a warm bench or staging in the greenhouse or a similarly bright place.

This grass looks wonderful in long, wending waves, and one mature plant can yield enough infants to work in a really exciting and innovative scheme. The variety 'Tatra Gold' makes fountains of bright lime green in spring followed by quivering inflorescences of silvery pink that blend beautifully with other late-spring flowers in the same colour range, like the enigmatic *Papaver orientale* 'Patty's Plum' or the

DRAMATIC CHANGE
Splitting grasses like miscanthus can be bicep building, although the end results are often very compact!

Quite tiny pieces of any grass, like this deschampsia, will grow into strong plants, as long as they have the bare essentials: a few new shoots and some decent roots.

mysterious hooded flowers and dusky foliage of *Lamium orvala*.

The bigger, tougher grasses, such as *Miscanthus sinensis*, can be treated in the same way, but may require a lot more effort to pull apart. Both miscanthus and forms of *Molinia* x *arundinacea* make big, dense clumps, that are almost impenetrable if you leave them too long. It's impossible to divide them by detaching pieces while they are still in the ground. The whole clump must be lifted, though sometimes this may mean asking for help. Two people on either side of even a huge clump should have it out in no time, and the whole co-operative experience should cement the friendship rather

than ruin it, especially if the second party gets a fair share of the proceeds. Once it is out, divide and sub-divide the clump until the whole thing is no more than a series of small crowns with a couple of growing tips and a few vital, white roots. At this stage it is difficult to believe that this tiny morsel will grow into a great big grass. Although the clump may be broken down into bigger chunks that can be planted out or potted up, one of the objectives when dividing grasses is to get rid of 'thatch' – old, dead stems that build up inside the clump – and it is only feasible to do that satisfactorily with smaller pieces. Water well when replanting whether in the ground or in pots.

Division

The big picture

This fine array of *Miscanthus sinensis* 'Flamingo' currently resides in big clay pots on the front terrace, where it makes a see through screen. It changes week by week, as the pink silky tassels become fluffy heads. All these plants came from one clump. Plants thrive by being divided; it gives them extra 'voomph'. And it gives you, as a gardener, lots of material to indulge your dreams.

OFFSETS

There are some plants that almost seem to laugh at you when you inspect them with an eye to propagating them. 'Why bother?' they seem to say 'We don't need your help, we can do it ourselves.' Without any intervention on your part, new plants simply appear spontaneously – and very often symmetrically – around the perimeter of the parent plant. Where there was originally just a sun, suddenly it is surrounded by satellites. The crowns of these plants naturally make offsets as an insurance policy in case the primary shoot gives up the ghost – mountain scree plants are typical of the types that colonize by offsets. Gardeners can exploit this characteristic to make new plants.

Echeveria and sempervivums produce large numbers of smaller plants around their skirts. The usual way to increase them is to simply pull off the rosettes that appear around the edge, pot them up and plant them out when the time comes. They don't even need pots: if they are pushed into the ground with a bit of their umbilical stem attached they will almost certainly take root. In some cases they don't even need soil.

Growing from offsets is the acknowledged method to increase all those succulents and similar plants that produce large numbers of smaller plants around their skirts. Although this can be done at any time of the year, the plants themselves usually dictate when they are ready to be increased. It is probably the most successful policy to detach them when they have reached a fair size – not as big as their parent plant, but big enough to survive on their own.

But other plants besides these succulents can also be split at their base without resorting to knives. Herbaceous lobelias have become a familiar sight in garden centres. Some of them can be grown from seed, but this is always a painstaking business since the seed is like dust and slow to germinate. A far speedier alternative, even though it may produce fewer plants, is simply to prise the plant apart into separate rooted rosettes, treating each piece as a rooted cutting and potting them up individually.

Another advantage of removing offsets is that this is a vegetative method of propagation. This means that, in effect, you are always 'cloning' the plant, and identical offspring will be produced whereas with seed there is no such guarantee. For coloured leaves and unusual forms, this is crucial.

193

Succulent rosettes

Sempervivums aren't just tough, they're really hard; their only needs are full sun and sharp drainage.

Often called houseleeks, sempervivums were traditionally grown between roof tiles or slates where they creep along filling spaces with their neat rosettes, existing only on any detritus that is washed down. Jovibarbas are a close relation and spread in the same way, making a cluster of offsets around the base at the end of long stoloniferous stems, which can be left to make a cushion or pulled off and grown on. Their name means literally 'Jove's beard' and this whole group has been associated with Jove or Jupiter; having them on your roof was meant to ward off lightning, Jupiter's thunderbolts. A new twist to this is the development of roof matting that consists of sedums and sempervivums woven into fabric. These green roofs are self-regenerating, long-lasting, and ecologically friendly, even more so if you grow your own plants – and presumably they are never struck by lightning!

Individual rosettes can be pulled away from the parent plant with the stem that attaches them to it. Dibble a hole in a pot of gritty compost and push the stem into it so that the new rosette is sitting on the surface of the soil. If you are hoping to make lots of new plants, use a module or cell tray and put one rosette into each compartment. When plants have rooted well, they can be potted on or planted out.

Sempervivums are tough and hardy, but echeveria are tender. Echeveria often start to collapse outwards of their own volition, and at this stage small rosettes can be pulled away from their parent and potted up individually. Since they are tender they should spend the winter in the greenhouse. By the next spring they will have multiplied again and the process can be repeated. Small wonder they are used so extensively in municipal carpet bedding schemes.

WINNING ROSETTES
ABOVE Echeveria are a delight for summer bedding, or just to admire in a decorative pot. Establish them first by detaching offsets and pushing each one into a plastic pot or module.
OPPOSITE Sempervivums, or houseleeks, have a plethora of different colours and texture. Propagating them is as easy as pie.

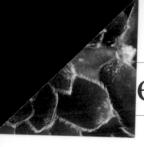

ed stems

Taking offsets is the only way to make more of your precious auriculas and the plants almost do it themselves.

Primula auricula is closely related to other members of the primula clan, but its demeanour and characteristics separate it from its shade-loving cousins, whose object in life is to imbibe the moisture and nutrients in the first few inches of soil. *Primula auricula*, a striking alpine plant with mealy leaves and a head of yellow flowers with white centres, has very different requirements.

Most primulas have a fibrous root system, but the roots of *P. auricula* are strong, thick and fleshy. They are designed by nature to help them penetrate crevices in the rock in search of water and nutrients in its stony habitat, high up in the mountains from Serbia to Northern Italy. In the depths of winter the thick leaves wrap themselves around one another, keeping the infant flower buds well protected from ice and snow. In the baking heat of summer they act as a reservoir, storing water for emergencies. Cultivated auriculas are a highly bred race of plants, but at heart they share the tenacity and hardiness of the wild auricula and the other related species, which were used in their development. No matter how much we refine their cultivation, to make them feel at home we need to remember where their forebears come from.

They are sometimes grown from seed, and breeders work tirelessly to make new cultivars, but when an exciting individual appears the only way to make more of it is to propagate vegetatively. Any named variety of these very desirable plants worth perpetuating can only be kept going by pulling off rooted offsets, potting them up, nurturing them and passing them on. Auricula addicts usually buy named plants that have already been bred, selected and increased from offsets of the original plant. Some of these old named varieties may have been propagated in this way for decades, sometimes even centuries. Until the advent of micropropagation there was no other method of increase that guaranteed the identity of a one-off.

We used to grow an auricula called 'George Swinford's Leathercoat', a gentle tawny-flowered plant, thick with farina, that dated back to the 18th century and must have been

ROOTED AURICULAS
Auriculas will make spontaneous offsets. These can be pulled from the plant, plunged into plugs and eventually potted up individually.

CLOSE TO NATURE
'Old Mustard' is an old variety and as close as can be to the wild auricula. It is a robust plant and really loves the outdoor life.

You can do this any time of the year, but the best time is probably just after flowering when there is plenty of light and warmth to help new plants grow.

Select a healthy plant in a pot or in the garden. Expose the base of the side shoots and, if possible, choose one that has already made roots; if none has, select a new healthy shoot. Sometimes an older plant lifts its crown proud of the soil or compost. To encourage root formation on the shoots, repot it in a bigger pot with its stems submerged under the soil, ensuring that the rosettes are flush with the surface of the soil; and come back for the shoots when they've had some months to make roots. Gently prise the shoot from the parent without damaging any roots; use a sharp knife if necessary. Carefully remove the older outer leaves by pulling them away cleanly from the main stem.

Trim the roots to about 5cm (2in) if necessary, so that they will not be bent, and plant in a clay pot with crocks in the bottom for good drainage and open, loamy compost. The top rosette should be level with the compost. Finish with coarse grit, water thoroughly and place outside or in a cool part of the greenhouse. If newly rooted offsets start to bloom, nip out the flowers so the plant puts its energy into making good roots and solid rosettes of leaves. When roots start to emerge from the drainage holes, pot on plants for 'show' or delicate 'fancy' varieties; the hardy border varieties can be planted out in the garden. Although show and fancy varieties must be kept under glass when flowering, to protect the colours and markings, they are hardy plants and both offsets and established plants benefit from a summer outdoors.

propagated from offsets hundreds of times before we were lucky enough to grow it. It was not a strong plant: vegetative propagation passes on not only genes, but also virus and other ailments. Nowadays many auriculas and other special members of the primula clan are produced by tissue culture. For the amateur gardener who wants to make more of a favourite plant, however, taking offsets is the only realistic option, and it's easy. The central rosette of leaves is formed on a thick stem, which makes new shoots at the very base that eventually form roots, and we can exploit this by removing offsets and making more.

Herbaceous offsets

If you want to make exact replicas of any cultivar or species, pulling off rooted rosettes is the best way.

We have a *Francoa sonchifolia*, a Devon plant originally from Langtrees at Croyde, and named 'Rogerson's Form' after its owner, Dr. Rogerson. It produces seed, but it is highly unlikely to come true. The only way to guarantee the same burnished, crinkly leaves and the solid spikes of deep magenta flowers is to propagate from offsets. Once potted, they come on speedily, making chunky plants in a few months. It is best to keep them in a coldframe or cold greenhouse and plant them out in late spring. This francoa is the hardiest of the bunch, but can be disfigured or occasionally killed stone dead in bitter weather, which is all the more reason to keep a few plants in reserve.

Heucheras are closely related to francoas and can be propagated in just the same way. They too are evergreen perennials, but though cold weather may make them look dishevelled, it seldom does them any permanent damage. Plants tend to lift themselves above the ground, which can be unnerving when you thought you were growing a ground-hugging rosette. On a large plant with numerous crowns, it is worth digging up the whole plant, breaking or cutting off several rosettes even if they have not yet made root, and potting them up or plunging each one into a cell tray of open, gritty compost. The parent plant can be returned to the garden, planting it more deeply so the remaining rosettes are level with the soil.

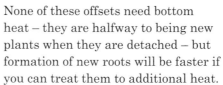

None of these offsets need bottom heat – they are halfway to being new plants when they are detached – but formation of new roots will be faster if you can treat them to additional heat.

Francoa, heuchera and some tiarella varieties (all members of the saxifrage family), have thick stems below their rosettes that, when pulled away from the parent plant, are sometimes rooted and sometimes rootless. Whatever stage they are at when they are detached, they will root firmly when plunged into compost, making a network of fine fibrous roots. Remove any old leaves to minimize the chance of rotting, water sparingly to keep the compost just moist and pot them on as soon as they fill their pots or cell trays with roots.

ONE TIMES SIX
One francoa may yield six new plants and still leave you with the original, central rosette.

FERNS

ANCIENT SURVIVORS
Ferns are among the most ancient plants on the planet, and the most fascinating. They have no flowers or seed, but all can be propagated from spores. Though it takes some time, it is magical.

In one quiet corner of our garden, beneath a line of tall beeches probably planted long ago as a farm hedge, the only colours are brown and green. The hedge runs east-west across half our south-facing slope, so when the canopy reaches out in summer, there is little chance of growing anything that flowers – there just isn't enough light. Then, every autumn, the brown of earth and leaf mould is enlivened by the russets and tawny orange of the beeches' fallen leaves. Many of us have this sort of dark corner in our gardens: it might be in the shadow of a garage or shed or the house itself, or of trees, hedges or fences, but whatever makes the shade, it is often seen as a problem. In fact it represents an exciting opportunity to grow some of the most beautiful and enduring plants in the world.

Ferns of all descriptions thrive in such corners, but from late autumn onwards the evergreen varieties come into their own. They persist through the winter and, provided they have some shelter from the coldest winds, their fronds will look as pristine in the early new year as they did in midsummer. Combining together perfectly, the variety of their individual fronds, from solid and shiny to soft and lacy, adds a wealth of textural interest.

It is a privilege to enter the domain of these strange, esoteric plants. Devon was Mecca for the pteridomaniacs, or fern collectors, of the Victorian era, and despite their depredations, the steep Devon banks still drip with polypody and asplenium, ooze dryopteris and polystichum. In this part of the world you are constantly aware of their presence and their indomitability. Just a couple of weeks after the flails have reduced the hedgerows to a brown crew-cut, polypody and lady fern, male ferns and hart's tongue reassert their verdancy and their lust for life.

Polypody is almost indestructible, and as 'at home' behind the wheelie bins of a city garden as it is in a rural wood. The species have bright green, simple fronds and spread readily; the strong, wandering roots binding loose soil. There are selections, like 'Cornubiense' with its more finely cut fronds, but polypody is, at heart, a straightforward plant valued for its ability to thrive cheerfully in any circumstances, rather than to provide lacy distractions. Leave such frippery to the soft shield fern, *Polystichum setiferum*, one of the most graceful and refined of our native ferns. This is positively frothy, especially in its forms 'Divisilobum' and 'Densum' in softest green with shaggy, rusty stems.

There are native evergreen ferns a plenty, but there are many exotic species that should be made to feel at home. Of them all, *Dryopteris erythrosora* from China and Japan is especially welcome for its graceful polished fronds, which are rich orange and ginger when they first unfurl.

Native or exotic, all these ferns can be propagated if we are prepared to take the time. Ferns are among the oldest plants on earth. Their ancestry stretches back 500 million years; in comparison, flowering plants have only just tipped up on the botanic scene. Although ferns have evolved during this time (there are as many as 10,000 species, only 50 of which occur in the British Isles), their sex-life has remained unchanged throughout. And a very esoteric sex-life it is.

Our familiar flowering plants reproduce each generation from seed, but ferns have no flowers and set no seed. They use spores to procreate, and there are two very distinct stages in this fascinating ritual. The first is the frond stage, where the fern spontaneously produces spores and releases them from the reverse of the fronds. Although we tend to think of spores as roughly like seeds, they are not – no sex is involved in their production.

The sex happens in the second stage, the prothallus stage. In moist conditions, the spores grow into small,

scale-like growths, which in turn produce male and female cells on their surface. The male cells move around in the water on the surface of the prothallus until they unite with the female cells: it is their union that creates the new fern. This dependency on water in the reproductive cycle has led scientists to believe that ferns are descended from aquatic plants.

All manner of magical properties have been attributed to ferns, mainly because before the invention of the microscope, nobody understood how they reproduced. Folklore had it that spores could bestow the gift of invisibility to those who believed in their magic power. Now that we know what really happens, we may no longer believe in these fairy stories, but one myth that persists is that the whole business of growing new ferns from spores is wreathed in mystery and fraught with problems. Although it takes time, and therefore patience, it is an exciting and rewarding process and full of an ancient poetry.

Propagating ferns takes a little time and trouble. It is a wonderful reminder that gardening is not a question of turning out a product, but rather of engaging in an ongoing process. In the case of ferns, the process had been going on for hundreds of millions of years, long before we were around, let alone aware of it.

Growing from spores

This emulates the way ferns have reproduced themselves for millions of years.

Spores can be collected from early summer through until early autumn. There are no hard and fast rules: watch the plants and check the reverse of the fronds. Some spores are green, and others brown, but as they ripen they generally darken and the little bead-like clusters become shaggy. Some of our native ferns set spores prolifically – sometimes the ground beneath them is covered with their spores, and any spiders webs suspended nearby take on a fine brown film where the spores have been caught up.

Sever a whole frond with scissors or secateurs at its base. The frond can be either plunged into a large paper bag if you want to sow its spores later, or laid on a piece of cartridge paper. The latter is more fun, because after a few days on a warm windowsill you can actually see a print of the fern frond on the paper, created by the fallen spores. Fold the paper in half or cover it with another piece to contain the spores, which can then be stored until the following spring or sown promptly.

Almost everyone who grows ferns from spores has their own special recipe for the compost. Some people use straightforward John Innes potting compost, but this does contain some fertilizer and ideally fern compost should contain no fertilizer should be sterile. Our mix here was equal parts sterilized loam and leaf mould, with some coarse sand added and crushed charcoal incorporated to prevent stagnation and help keep the compost

sweet. Fill a clay pot or clay pan with the compost, leaving a 'breathing space' of about 1cm (½in) between the top of the firmed compost and the top of the pot. To sterilize thoroughly, cut out a

HART'S TONGUE
The spores on the back of *Asplenium scolopendrium* leaves are like low mounds of rusty velvet.

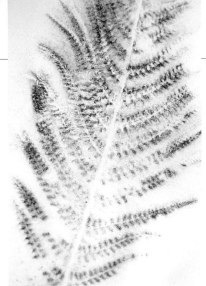

of the brick dust, either straight from the paper or on the end of a knife blade. Replace the glass. It will stay there until the fernlets are separated, keeping out foreign bodies and maintaining moisture levels – vitally important if fertilization is to take place. The most effective way to do this is to place your pan in a shallow dish of rainwater or distilled water and keep on topping it up. If it dries out, that will put paid to the process. Place the whole caboodle in a warm, shady place. After several weeks a green film should appear on the surface of the compost. This is made up of the developing prothalli, the inbetween stage. Eventually tiny fronds will begin to appear. The glass can be removed at this stage, although a warm, moist atmosphere must be maintained around the developing plants.

When they are large enough, they can be carefully lifted in small batches, separated, and pricked out into seed trays – a fiddle, but immensely satisfying, process. Make sure that your compost is well-watered and drained before starting this delicate operation. Later, you can pot the plants individually into small clay pots in loam-based compost. Top-dress the pots with grit to keep the surface cool and to ensure good drainage and even water distribution. Ferns are essentially shade lovers, so always ensure that young ferns are kept out of direct sunlight.

Plant out hardy varieties into the open garden when they are big enough to fend for themselves. Pot on tender species and keep them undercover – although many are perfectly happy having a summer vacation outside in a damp, shady spot.

COLLECTING SPORES
It's great fun to lay a frond on a piece of white paper and days later lift it to see its impression. Fold the paper and carefully spread the spores over the surface of your compost.

circle of clean paper to cover the top of the compost in the pot, place it on the surface. Pour boiling water onto the paper. Keep on pouring until the whole pot feels hot and the compost is saturated. Take off the paper and cover it with a piece of clean glass. Allow it to drain and cool, then cover the surface of the compost with a fine layer of brick dust to provide a sterile surface.

Tap the spores into the fold of the paper on which they have been collected and gently sow them as evenly as possible onto the surface

Growing from bulbils

A few ferns will form bulbils along the midribs of their fronds. Sometimes these develop spontaneously into new plants, but sometimes they need encouragement.

FERTILE FROND
A few ferns make clones of themselves along their midrib. If the whole frond is laid down on compost, each one may grow into a separate plant.

Polystichum and *Asplenium* species are good examples of this habit. For the ferns, bulbils on the fronds are a good back-up alternative to their intricate sexual reproduction. For the gardener they are another opportunity to increase our stocks.

A frond can develop a whole series of separate plantlets along the midrib, so that the classic fern form of the frond is barely distinguishable. At this stage, the whole frond can be lowered and laid on top of compost in a seed tray or large pot; if the plantlets are already well developed, the frond can be severed from the plant.

To ensure that the base of each of the new fernlets is in contact with the compost, use short pieces of florist's wire or staples to pin them down. Water well and keep in a warm, shady place, frequently spraying with a fine mist to maintain moisture levels. After a few months, each of the little ferns will have made its own roots, and at this stage they can be separated and potted on individually.

Some ferns make only one new bulbil per frond. In this case the frond can be bent over pushing the new fernlet and its bulbil gently, but firmly, into a pot of compost. Again it can be held in place with a staple. The mother plant, together with its satellite pots, should be kept in a moist, warm, shady place and the new plant can be cut from the frond once rooted, launching it on an independent existence.

DIVIDING A FERN
OPPOSITE *Onoclea sensibilis* is a wandering fern that lends itself to division. A few months later, these pieces will be ready for planting out.

SINGLE FERNLET
This perfect little fern can be detached and potted, or pinned down and given time to grow on.

NEW PLANTS FROM DIVISION

Ferns have very fibrous roots. Large clumps of established ferns that form separate crowns, such as the male fern *Dryopteris filix-mas* and the lady fern *Athyrium filix-femina*, can be lifted and separated in exactly the same way as clump-forming perennials (see pp.170-1). Cut back the fronds in early spring, lift them and either pull them apart with your hands or, if that's too tough a proposition, use two garden forks back to back (see p.171).

Break up into chunks, each with at least one crown and adequate roots, and replant directly, adding leaf mould or home-made compost.

Some ferns have running, stoloniferous roots, and this habit can be exploited to make new plants. *Onoclea sensibilis*, the sensitive fern, and *Matteuccia struthiopteris*, the shuttlecock fern, are two good examples. Simply sever the stolon, which joins one plant to the next, lift the furthest plant and pot it up individually. Water well, allow to establish and plant out when well rooted. Alternatively the offsets can be planted out directly if they have enough roots to sustain themselves independently.

AFTERCARE

Gardening is addictive. One of the sure signs that you are well on the way to being hooked is an almost overwhelming desire to make ever more new plants.

Quite apart from satisfying the cravings, there are numerous good reasons for propagating your own plants. When you have grown a plant from a seed or cutting, nurtured it during its infant stages and put it into the garden as a young adult, you know it intimately. Nothing puts the gardener in closer touch with the earth and its growing cycles than rearing their own plants from scratch.

There are, of course, more prosaic reasons for growing new plants too – and for hanging onto those you already have. Plants are expensive. Without a large cash surplus with which to buy an 'instant' garden, your ambitious plans for drifts of this over here and swathes of that over there may never become reality unless you grow your own stock of plants.

Knowing how to coax seeds into germinating or where to sever a piece of your favourite shrub to make a successful cutting comes with experience, but nurturing the new plants through infancy, childhood and adolescence, until they can take their place in a grown-up garden, is just as crucial. Whatever method we use to propagate our plants, the stages that follow are just as vital as that first act. Once we have initiated the process, we have a duty of care to the young plant. Furthermore, and leaving any sentiment aside, if we want strong, healthy plants for our garden we need to give our new charges the best possible aftercare.

It is vital to make a careful analysis of what is needed to offer optimum conditions to the plants you want to grow and conserve. What do you really need? What can you do without?

A FLYING START

Giving your new plants the best environment in which to thrive is imperative. As a general rule, extra warmth in the form of a heated propagator (or if you are taking the whole thing very seriously, a heated bench with soil-warming cables) will speed up the processes of germination and rooting, but this is not true in every case. Some seeds are happiest in cooler conditions, so bottom heat isn't necessary for all and sundry; after all, gardeners have been propagating plants for a long time without it.

Once seed has germinated, it must be pricked out and given its own space

FROM THE ROOTS
Ideally a garden is self-sustaining in every way. Dead plants are composted to feed living ones, and older plants give material for new ones. Of course you'll want to add different plants; how thrilling it is when the plants you use are ones you have grown yourself.

as soon as it has its first true leaves and a well-formed root system, before the seed tray becomes overcrowded. Leave it too long and the seedlings start vying with each other for resources – nutrients, water and light. The same goes for cuttings made from stems or roots, for the small plantlets that form on leaf cuttings, or the young corms or bulbs that emerge as a result of our intervention.

POTTING ON AND PLANTING OUT

Young plants, just like young animals, grow very rapidly. And just as rapidly, their appetite increases, they get thirsty faster and they need more space too.

As plants grow they need more nutrients to ensure that their stems grow strong, their root systems develop and their leaves have what they need to stay green and keep growing. Though seedlings may start life in a rather bland compost, as they grow they will need stronger stuff and should be potted on regularly into bigger pots and richer compost.

When you come to plant them out, the same ground rules apply to your home-produced young plants as they would to any other young plants. Weed the ground thoroughly, especially to clear perennial weeds – docks, couch grass and bindweed – and after you introduce your plants try to ensure there isn't any competition from weeds or any other boisterous border inhabitants.

Prepare the ground well by turning it over gently with a fork and incorporating generous quantities of home made compost. If you haven't enough yet, a commercial organic 'soil improver' is a good alternative. Water plants well even before you plant, and water the planting hole to ensure moisture isn't sucked away from your plants. After planting, water again to settle the soil and close any larger pockets of air. Juveniles cannot yet take a rich, adult diet, but as they begin to settle in you can give them the occasional treat of a very dilute organic liquid fertilizer.

This is all that is needed for cuttings that were pushed into the open ground or raised outside, and seeds that were sown directly or in a seedbed, but these are likely to be only one part of your propagating efforts. For the rest, investing a bit more care will bring abundant returns.

PROVIDE PROTECTION

Even when they are rooted and potted up, young plants need help and care to get them through their first precarious weeks and months. There is nothing quite so discouraging as losing plants that you have helped into life. One harsh winter, soon after we first started the nursery here, more than half the cuttings we had successfully established were killed by a combination of wet and cold. There was no room left in the small coldframe we'd built and, being the over-optimistic beginners we were, we assumed that once they were rooted our cuttings would survive without too much mollycoddling. Many were cuttings from hardy shrubs, after all. But being such young plants, they had no resources to fall back on: no ripe wood and no mature root systems. With no protection, they succumbed to the cold and wet.

Providing protection from the elements and maximizing available light and heat will lengthen the growing season, and growing under glass also enormously broadens the scope of what can be grown.

COLDFRAMES

Most hardy plants, even in their infancy, need no more than to be kept above freezing and given the benefit of full daylight. A simple coldframe sited in a light and open place is enough for them. If there are emergency cold nights to cope with, you can add insulation for extra protection.

Coldframes have fallen from fashion, but they offer a realistic alternative for many of us – especially when space is limited. Because they are at ground level, they must be sited to

take advantage of all available light, and the glass must be kept clean. Heat builds up fast in coldframes, so efficient ventilation is vital. Dutch lights or other tops should be removed in summer, and even on warm spring days they should be raised to prevent overheating; whitening the glass with shading wash will prevent scorch. In cold weather, insulation can be added to maintain frost-free conditions; commercial bubble-wrap is cheap and long-lived.

GROWING IN A GREENHOUSE

Small greenhouses provide protection and a space to work. Aluminium-framed greenhouses are the least expensive option, and putting them up yourself is fairly straightforward.

ALL-YEAR ASSET
A greenhouse is busiest in the spring, when full of early sowings, but at the other end of the year it is just as useful for tender cuttings.

Their disadvantage is that they rapidly heat up in summer; plants can shrivel and gardeners find it uncomfortable to work, even with shading wash or blinds. Conversely, aluminium greenhouses quickly become fridge-like in the winter. Lining the glazing with bubble wrap can help, but this may still fail to provide frost-free conditions for young or tender plants during cold spells. Even if heaters are used, energy costs can outweigh the value of the plants. Heat and cold pass through a single pane of glass easily, and metal frames exacerbate the process.

Temperature fluctuations are less extreme in timber-framed greenhouses, and heat loss is significantly reduced in those with half-timber sides. Wood is an efficient insulator. Half-brick greenhouses store heat and maintain the most even temperatures. Size is important too: the bigger the volume of air, the slower it loses heat. A lean-to greenhouse that makes use of a sunny house wall is very efficient, and usually has a pleasant atmosphere.

Apart from the temperature, the other major factor that must be considered when growing under glass is ventilation. A buoyant atmosphere with a constant flow of fresh air is an all-important safeguard against fungal diseases, which are the main problem for new and over-wintering plants. Louvred vents and windows and sliding doors that can be pulled back easily should be adequate. With good built-in ventilation, there should be no need for electric fans.

Is there any need for electricity? The keener you are, the greater the need. Darkness descends far too early during the main propagating season of early spring, and again during the late autumn and winter months, which are the optimum times for taking root cuttings and for nurturing tender plants. It may be dark by the time many gardeners are home from work, and we may all be too busy attending to the garden outside to get the chance to work in our greenhouses during daylight hours. If your greenhouse is close to the house, installing electricity is fairly simple and not too expensive – bulkhead lights suitable for outdoor use are cheap. We garden organically at Glebe Cottage and though there are few aphids problems, slugs and snails can decimate trays of seedlings. It's worth regularly checking the trays.

Electricity also increases scope for propagation exponentially. Soil-warming cables can be laid in a simple timber frame filled with sand to provide heat to germinate seed and enable cuttings to put down roots rapidly. If this is too ambitious, even a small propagating case that accommodates a few half trays expedites all propagation. On colder nights, electricity gives you the opportunity to use a small fan heater on a frost-protection setting.

HARDENING OFF

The concept of 'weaning' plants is just as applicable to plants as it is to young creatures. As plants grow stronger, you need to help them acclimatize to tougher conditions in preparation for their outdoor life. As conditions outside warm up in spring, your plants will need to be 'hardened off'.

The phrase is really self-explanatory: as young plants move into the conditions they will experience when they are finally planted out, their soft growth will literally toughen up. Hardening off involves opening up the greenhouse so that the temperatures within and without are closer, or opening the glass 'lights' or plastic covers on coldframes containing young plants, and returning the plants lifted out by day. In the evening as temperatures fall, return the plants to the frame and close it, or close your greenhouse up, so that young plants are gradually introduced to the environment they will have to contend with.

YOUR GROWING NEEDS MAY GROW

Whichever sort of greenhouse or coldframe you decide on, chances are it will seem inadequate after a couple of years, especially if the propagating bug takes hold. The wider the range of plants, the larger the space needed. Space used for bringing on seeds and cuttings during spring and summer can be turned over to protecting tender echeverias and aeoniums or rooting cuttings from tender perennials in winter. Working out an approximate timetable is always a good idea if you want to make maximum use of your glass. Start from the results you want to achieve, rather than what can be accommodated initially. Plants grow

fast, and what started out as a couple of trays of seedlings may end up occupying a large area of staging when pricked out into bigger pots.

In an ideal world, a perfect solution is a small timber-framed greenhouse that can be extended to meet growing needs. For the rest of us realism must prevail. Fit your choice of building or frame to the amount of time you are able to devote to nurturing the plants you start. That time is critical: given the same time, similar success rates will be achieved whichever route you choose, and you will experience the same thrill whether your new plants are growing on a windowsill, in a cold frame or in a greenhouse.

RARING TO GO
When winter is over, the covers come off and the greenhouse doors are opened wide, revealing scores of healthy young plants ready to get their roots into the summer quarters in the wider garden.

Matching plants to techniques

Hopefully you have already been bitten by the propagating bug and you are itching to get started. Increasing your plants is a joy, but sometimes you hesitate about which method will work best – how do you turn your solitary salvia into a whole swathe? Experiment is vital and experience is the best teacher but it's still reassuring to have a few extra pointers.

The tables on the next pages give some indications of which methods will work for different plants. In many cases there is more than one alternative so you can see which suits you best or, where you have always grown something from seed, you can have a go at a new technique – root cuttings perhaps or chopping up leaves.

When you consider how many desirable plants there are in the world, there is no way this list can pretend to be comprehensive, on the other hand it is a sound starting point. If the plant you want more of is missing, try to find something close to it in the tables and have a go – even if you are unsuccessful all you are losing is one opportunity, you can try again with another method. There are many other sources you can investigate for further or more specialised information.

The more you make your own plants and grow your own garden, the more familiar you become with which methods are most appropriate for which plants. Half the fun is finding out and setting yourself challenges to see if you really can put roots on this bit of stem or persuade these seeds to germinate. Your repertoire will increase and there is a huge buzz to be had from developing your own modifications to established methods and passing them on to like-minded friends – it's just as much fun as sharing the plants you help into life.

SHRUBS AND SUBSHRUBS

S = seed (for species) L = layer SW = softwood cutting
SR = semi-ripe cutting HW = hardwood cutting
R = rooted suckers

	S	L	SW	SR	HW	R
Acer (some)	●					
Abelia			●	●		
Abutilon	●		●	●		
Acacia	●					
Aloysia			●			
Amelanchier	●	●				
Aronia	●		●	●		●
Artemisia			●	●		
Aucuba				●		
Berberis				●		
Brachyglottis			●	●		
Buddleja			●	●	●	
Buxus			●	●		
Callicarpa			●			
Callistemon	●			●		
Calluna		●	●			
Camellia	●		●	●	●	
Caragana	●	●	●	●		
Carpenteria	●			●		
Ceanothus		●	●	●		
Ceratostigma			●	●		●
Chaenomeles			●	●		●
Chimonanthus	●	●	●			
Choisya			●	●		
Cistus	●		●	●		
Clerodendrum	●		●	●		●
Clethra	●		●	●		
Colutea	●		●			
Cornus	●	●	●	●	●	
Coronilla			●	●		
Corylopsis	●	●				
Corylus	●	●				●
Cotinus			●	●		●
Cotoneaster	●	●		●		●
Crinodendron	●		●	●		
Cuphea	●		●			
Cytisus	●		●	●		
Daphne	●	●	●	●		
Dendromecon	●		●			●
Deutzia			●	●		
Dipelta			●			
Dracaena		●	●	●		
Drimys	●		●	●		
Elaeagnus	●			●	●	●
Enkianthus	●			●		
Erica	●	●	●			
Erysimum			●			
Erythrina	●		●	●		
Escallonia			●	●		
Euonymus	●		●	●		
Exochorda x macranthus	●	●				
	S	L	SW	SR	HW	R

	S	L	SW	SR	HW	R
x Fatshedera			●	●		●
Fatsia			●	●		
Forsythia			●	●	●	
Fremontodendron				●		
Fuchsia	●		●	●		●
Garrya				●		
Gaultheria		●		●		●
Genista	●		●	●		
Griselinia	●		●	●		
Halimium			●	●		
Hamamelis	●					
Hebe			●	●		
Heliotropium	●		●	●		
Hibiscus	●		●	●		
Hippophäe	●			●		●
Hydrangea			●	●		
Hypericum	●		●	●		
Hyssopus	●		●	●		
Indigofera	●		●	●		
Ilex	●	●		●	●	
Itea			●	●		
Jasminum		●	●	●		●
Kalmia	●	●	●			
Kerria			●	●		●
Kolkwitzia			●	●		
Lantana	●		●	●		
Lavandula	●		●	●		
Lavatera	●		●			
Leptospermum			●	●		●
Leycesteria	●		●			●
Ligustrum			●	●		
Lonicera	●	●	●	●	●	●
Magnolia	●					
Mahonia	●					●
Myrtus	●		●	●		
Olearia	●		●	●		
Osmanthus			●	●		
Paeonia	●					
Perovskia			●	●		
Philadelphus			●	●		
Phlomis	●		●	●		
Photinia				●		
Physocarpus	●		●	●	●	
Pieris		●	●	●	●	●
Pittosporum	●			●		
Potentilla			●	●		
Prostanthera	●		●	●		
Prunus (shrubby types)						●
Pyracantha			●	●		
Rhamnus			●	●		
Rhododendron	●	●	●			
Rhus			●	●		●
Ribes		●	●	●	●	●
Robinia	●					●
Romneya	●		●			●
Rosa	●	●	●	●	●	●
Rubus	●	●	●	●		●
	S	L	SW	SR	HW	R

	S	L	SW	SR	HW	R
Salix		●	●	●	●	●
Sambucus			●	●	●	
Santolina			●	●		
Sarcococca	●		●	●		
Skimmia				●		
Solanum		●	●	●		
Sophora	●					
Sorbaria			●			●
Spartium	●		●			
Spiraea			●	●		
Stephanandra			●	●		
Styrax			●	●		
Syringa						●
Tamarix				●	●	
Vaccinium	●			●		
Viburnum	●	●	●	●	●	
Vinca			●	●	●	
Weigela			●	●	●	
Yucca						●
	S	L	SW	SR	HW	R

CLIMBERS

S = seed (for species) L = layers B = basal cuttings
SW = softwood SR = semi-ripe cuttings HW = hardwood cuttings
D = division R = self-suckering roots

	S	L	B	SW	SR	HW	D	R
Actinidia	●	●		●	●			
Akebia	●				●			
Billardiera	●			●	●			
Bougainvillea		●			●			
Clematis	●	●		●	●			
Clianthus	●			●				
Hedera		●		●	●			
Hoya				●				
Humulus		●		●	●		●	
Hydrangea		●		●	●			
Ipomoea	●							
Jasminum				●	●			
Lathyrus	●						●	
Lonicera		●			●	●		
Parthenocissus				●	●			
Passiflora	●			●	●			
Plumbago				●	●			
Rosa	●			●	●			●
Schisandra	●			●	●			
Schizophragma	●			●	●			
Solanum				●				
Stephanotis	●			●	●			
Trachelospermum	●	●		●	●			
Tropaeolum	●		●					
Vitis						●		
Wisteria	●	●						
	S	L	B	SW	SR	HW	D	R

217

PERENNIALS

S = seed (for species) BC = basal cutting SC = stem cutting
LC = leaf cutting D = division R = root cuttings O = offsets

	S	BC	SC	LC	D	R	O
Acanthus	●					●	
Achillea	●				●		
Aconitum	●	●			●		
Actaea	●						
Adenophora	●	●			●		
Agastache	●		●				
Ajuga					●		●
Alcea	●						
Alchemilla	●				●		
Ammi majus	●						
Anaphalis					●		
Anchusa						●	
Anemone (Japanese types)					●	●	
Anemone nemorosa					●	●	
Anemone rivularis	●					●	
Anemone vitifolia	●						
Anemonella					●		
Anethum graveolens	●						
Angelica	●						
Anthemis		●	●				
Anthriscus	●						
Aponogeton	●				●		
Aquilegia	●						
Aralia			●				
Arenaria	●	●					
Artemisia			●				
Aruncus	●				●		
Asarum					●		
Aster		●	●				
Astilbe					●		
Astrantia	●				●		
Auricula	●				●		●
Bergenia					●		
Bidens	●				●		
Brunnera	●				●		
Calamintha	●		●				
Calendula	●						
Campanula	●	●			●		●
Cardamine	●			●	●		
Carex	●				●		
Carlina	●						
Catananche	●						
Centaurea	●				●		
Centranthus	●				●		
Cephalaria	●				●		
Chrysanthemum		●	●		●		
Cimicifuga	●				●		
Cirsium					●	●	
Convallaria					●		
Corydalis	●				●		●
Cosmos	●	●					
Crambe	●					●	
Cynara		●					●
Delphinium	●	●					
Dianthus	●		●				
Diascla			●				
Dicentra	●				●		
	S	BC	SC	LC	D	R	O

	S	BC	SC	LC	D	R	O
Dictamnus	●						
Dierama	●				●		
Digitalis	●						
Diplarrhena					●		
Doronicum	●				●		
Echinacea	●				●		
Echinops	●				●	●	
Epimedium	●				●		
Eremurus	●						●
Erigeron	●				●		
Erodium	●					●	
Eryngium	●					●	
Euphorbia	●	●			●		
Filipendula	●				●		
Foeniculum	●						
Francoa	●						●
Gaillardia	●				●		
Galium	●				●		
Gazania	●	●					
Geranium	●	●			●		
Geum	●				●		
Gunnera	●				●		●
Gypsophila	●	●					
Hedychium	●				●		
Helianthus	●				●		
Helleborus	●				●		
Hemerocallis	●				●		
Hesperis	●	●	●				
Heuchera	●	●			●		●
Hosta	●				●		
Inula	●				●		
Knautia	●			●			
Kniphofia	●				●		●
Lamium	●	●	●		●		
Liatris	●	●			●		
Ligularia	●				●		
Liriope	●						●
Lobelia	●				●		●
Lupinus	●	●					
Lychnis	●				●		
Lythrum	●				●		
Macleaya						●	
Malva	●		●				
Meconopsis	●						
Melianthus	●		●				
Mentha	●				●	●	
Mimulus	●				●		
Monarda	●	●			●	●	
Nepeta	●	●	●				
Oenothera	●				●		
Ophiopogon	●				●		●
Origanum	●		●		●		
Osteospermum	●		●				
Papaver orientale	●					●	
Pelargonium		●	●				
Penstemon	●	●	●				
Persicaria			●		●		
Phlomis (herbaceous)	●		●				
Phlox	●	●	●		●	●	
Phormium							●
Podophyllum	●				●		
Polemonium	●	●					
	S	BC	SC	LC	D	R	O

	S	BC	SC	LC	D	R	O
Polygonatum					●		
Potentilla	●				●		
Primula	●				●		
Pulmonaria	●				●	●	
Ranunculus	●				●		
Rodgersia	●				●		
Rudbeckia	●				●		
Salvia	●	●	●				
Sanguisorba	●	●			●		
Scabiosa	●				●		
Schizostylis	●				●		
Sedum	●	●	●	●	●		●
Semiaquilegia	●						
Sempervivum	●						●
Senecio	●				●		
Sidalcea	●	●			●		
Sisyrinchium	●				●		
Stachys	●	●			●		
Symphytum	●				●	●	
Tiarella	●			●	●		
Tolmela	●			●	●		
Tricyrtis	●				●		
Trillium	●				●		
Verbascum	●					●	
Verbena	●	●	●				
Veronica	●	●	●				
	S	BC	SC	LC	D	R	O

BULBS, CORMS, TUBERS AND RHIZOMES

S = seed (for species) W = wounding (scooping, scoring)
O = offsets S = scaling or 'chipping' TS = twin scaling
D = dividing clumps (including rice)

	S	W	O	S	TS	D
Agapanthus	●					●
Allium[1]	●		●			●
Amaryllis				●		●
Anemone						●
Arisaema	●					●
Arum	●					●
Asarum	●					●
Camassia	●		●			●
Canna						●
Cardiocrinum	●		●			
Chionodoxa	●		●			
Colchicum			●			●
Crinum	●		●			●
Crocosmia	●	●	●			●
Crocus	●		●			●
Cyclamen	●					
Dahlia[2]	●					●
Dracunculus						●
Eranthis	●					●
Erythronium	●					●
Eucomis[3]	●		●		●	
Freesia	●					●
Fritillaria[4]	●		●	●		●
	S	W	O	S	TS	D

	S	W	O	S	TS	D
Galanthus	●		●		●	●
Galtonia	●		●			●
Gladiolus	●	●	●			●
Hedychium						●
Hermodactylus[4]	●					●
Hyacinthoides	●				●	●
Hyacinthus		●			●	●
Iris	●					●
Ixia	●					●
Lachenalia			●			●
Leucojum	●					●
Lilium[1]	●			●	●	
Muscari	●		●			●
Narcissus	●		●		●	●
Nectaroscordum	●		●			●
Nerine	●		●			●
Ornithogalum	●					●
Paeonia	●					●
Puschkinia	●		●		●	●
Roscoea	●					●
Scilla	●		●			●
Sisyrinchium	●					●
Sparaxis	●					●
Sternbergia	●		●			●
Triteleia	●					●
Tulbaghia	●					●
Tulipa	●		●			●
Watsonia	●					●
Zantedeschia	●					●
	S	W	O	S	TS	D

[1] Allium and Lilium also make stem or flowerhead bulbils that can be removed and planted. [2] Also basal cuttings for dahlias. [3] Eucomis can also be propagated by leaf cuttings. [4] These plants make rice rather than decent-size bulblets

HOUSE AND CONSERVATORY PLANTS

S = seed (for species) LV = leaf vein LP = leaf petiole
LS = leaf slashing or squares BC = basal cuttings SC = stem cuttings
D = division R = root cuttings

	S	LV	LP	LS	BC	SC	D	R
Aeonium	●		●					
Begonia	●	●	●	●	●	●	●	
Brugmansia	●					●	●	
Chlorophytum							●	
Crassula			●			●		
Cyclamen	●							
Dieffenbachia							●	
Echeveria	●		●		●			
Gloxinia	●		●	●				
Impatiens	●					●		●
Pelargonium	●					●		
Plectranthus	●					●		
Saintpaulia		●	●		●			
Sansevieria		●						
Senecio	●					●		
Solenostemon	●					●		
Streptocarpus	●	●	●					
Tradescantia							●	●
	S	LV	LP	LS	BC	SC	D	R

Index

Note: page numbers in bold refer to information contained in captions.

Acknowledgements

Working on this book has been a joy. Everyone involved has been wholeheartedly enthusiastic about the project. It would not have happened without them.

Special thanks to Jonathan Buckley for his superb photography; our collaboration has been so rewarding and the Olympian standards he imposes on himself make us all strive to match them.

Lorna Russell, our editor, is simply the best; since the inception of the book she has worked tirelessly, selflessly and with passion to ensure it is as special as it could be.

We are all enormously grateful to *Gardens Illustrated* and in particular to its gem of an editor Juliet Roberts. This book has its roots in a series on propagation commissioned by *G.I.* and we are indebted to them for their support. The art director on that series was Sian Lewis, serendipitously she was able to work with us this time too and has made a brilliant job of it.

I have always loved growing plants but I owe an enormous debt of gratitude to all the friends and gardeners who have shown me how to propagate: my lovely friend Richard Lee who could put roots on a stick; Alan Street, a true inspiration and the best giggle imaginable; and Elizabeth Strongman, who showed me how to pollinate hellebores (when they eventually raised their heads from frozen ground) and inspired me to adventure and experiment. All these people are so passionate about plants and such good fun.

Last but not least the encouragement from my very wonderful family past and present.

And a special thanks to my two best propagating friends Tina, my penknife, and my favourite chopstick.

10 9 8 7 6 5 4 3 2 1
Published in 2010 by BBC Books, an imprint of Ebury Publishing.
A Random House Group Company

The Random House Group Limited Reg. No. 954009
Addresses for companies within the Random House Group can be found at
www.randomhouse.co.uk

A CIP catalogue record for this book is available from the British Library.

ISBN 978 1 846 07847 7

The Random House Group Limited supports the Forest Stewardship Council
(FSC), the leading international forest certification organisation. All our titles
that are printed on Greenpeace approved FSC certified paper carry the FSC logo.
Our paper procurement policy can be found at www.rbooks.co.uk/environment

COMMISSIONING EDITOR
Lorna Russell
COPY-EDITOR
Candida Frith-Macdonald
DESIGNER
Sian Lewis
PHOTOGRAPHER
Jonathan Buckley
PRODUCTION
Antony Heller
Colour origination by
AltaImage, London
Printed and bound by
Firmengruppe APPL, aprinta druck,
Wemding, Germany

To buy books by your favourite
authors and register for offers,
visit www.rbooks.co.uk